Friendship 7

The Epic Orbital Flight of John H. Glenn, Jr.

Other Springer-Praxis books
of related interest by Colin Burgess

NASA's Scientist-Astronauts
with David J. Shayler
2006
ISBN 978-0-387-21897-7

Animals in Space: From Research Rockets to the Space Shuttle
with Chris Dubbs
2007
ISBN 978-0-387-36053-9

The First Soviet Cosmonaut Team: Their Lives, Legacies and Historical Impact
with Rex Hall, M.B.E.
2009
ISBN 978-0-387-84823-5

Selecting the Mercury Seven: The Search for America's First Astronauts
2011
ISBN 978-1-4419-8404-3

Moon Bound: Choosing and Preparing NASA's Lunar Astronauts
2013
ISBN 978-1-4614-3854-0

Freedom 7: The Historic Flight of Alan B. Shepard, Jr.
2014
ISBN 978-3-3190-1155-4

Liberty Bell 7: The Suborbital Mercury Flight of Virgil I. Grissom
2014
ISBN 978-3-319-04390-6

Colin Burgess

Friendship 7

The Epic Orbital Flight of John H. Glenn, Jr.

 Springer

Published in association with
Praxis Publishing
Chichester, UK

Colin Burgess
Bangor, NSW, Australia

SPRINGER-PRAXIS BOOKS IN SPACE EXPLORATION

Springer Praxis Books
ISBN 978-3-319-15653-8   ISBN 978-3-319-15654-5   (eBook)
DOI 10.1007/978-3-319-15654-5

Library of Congress Control Number: 2015933843

Springer Cham Heidelberg New York Dordrecht London

Front cover: John Glenn is inserted into *Friendship 7* ahead of his MA-6 mission. (Photo: NASA)
Back cover: John Glenn suited up in 1962 prior to his Mercury flight (left) and thirty-six years later, for his flight aboard Space Shuttle *Discovery*. (Photos: NASA)
Cover design: Jim Wilkie
Project copy editor: David M. Harland

Printed on acid-free paper

Springer International Publishing AG Switzerland is part of Springer Science+Business Media (www.springer.com)

Contents

The legacy of heroes is the memory of a great name and the inheritance of a great example.

– Benjamin Disraeli, British politician and author (1804–1881)

A lot of people ask why a man is willing to risk hat, tail and gas mask on something like this space flight. I've got a theory about this. People are afraid of the future, or the unknown. If a man faces up to it and takes the dare of the future he can have some control over his destiny. That's an exciting idea to me, better than waiting with everybody else to see what's going to happen.

– Lt. Col. John H. Glenn, Jr., U.S. Marine Corps, NASA astronaut

Foreword

As the first American to orbit the Earth, John Glenn's accomplishments during his 20 February 1962 flight aboard *Friendship 7* are well documented and beautifully summarized in the pages that follow. The national prestige and sense of technological achievement were tremendous outcomes from his three-orbit flight. It got America firmly back in the Space Race with the Soviet Union.

One of the lesser known impacts of his flight was the effect it had in recruiting another generation of space explorers that would go on and follow in his footsteps. If you asked a hundred shuttle-era astronauts who or what event most influenced them to become an astronaut, you would find John Glenn and his *Friendship 7* flight at or near the top of the list. He is that iconic within the history of human space flight.

John Glenn with the All-Ohio crew of shuttle mission STS-70. From left: Don Thomas, Terry Henricks, Glenn, Mary Ellen Weber, Nancy Currie, and Kevin Kregel. (Photo courtesy of Don Thomas)

For me personally, watching Alan Shepard's launch on 5 May 1961 on a small black-and-white TV screen in my elementary school gymnasium first planted the seed of exploration in my head. Then the next year we all returned and watched America's newest hero, John Glenn, as he blasted off into space aboard a much larger and more capable Atlas rocket. He carried the dreams of many young students along with him that day, and I was right there with him.

After the launch we all marched single-file back to our classrooms. Fortunately my desk was in the back of the room along the windows which overlooked the school playground. Sitting in my seat I spent the rest of the morning not paying attention to my

Shuttle astronaut Don Thomas, Ph.D. (Photo: NASA)

first-grade teacher, but staring out the window. My eyes strained as I searched the sky hoping to see Glenn's *Friendship 7* capsule passing overhead. Little did I know at the time that his capsule was too small to ever be seen or that his orbital ground track was nowhere near Cleveland or the rest of my home state of Ohio, but that mattered little to me that day. My imagination ran wild with the thought that an American, and someone from my home state, was now orbiting the Earth. The evening paper ran the bold headline "Glenn Orbits the Earth and Safely Returns." In the following weeks *Life* magazine ran incredible stories and pictures from his flight. Everything I saw and read was awe inspiring.

One thing that particularly struck me was Glenn's descriptions of watching sunrises and sunsets and other views of the Earth. They were vivid which made me feel that I was right there with him inside *Friendship 7* looking out the window myself. But as good as the descriptions were, I craved more detail. I wanted to see the Earth and her glorious sunrises and sunsets with my own eyes. I knew I had to follow in Glenn's footsteps. While only six years old at the time, I was even more excited about becoming an astronaut myself and flying in space one day. And it is fair to say Glenn's *Friendship 7* mission had a similar impact on thousands – if not millions – of young students across the United States and around the world.

Thirty-two years later, I finally had the opportunity to reach space myself aboard Space Shuttle *Columbia* on the STS-65 mission, my first of four flights. Minutes after achieving orbit I floated to the window and got my first glimpse of the Earth, 175 miles below. I gasped and exclaimed, "Wow! How beautiful!" Even though I had read many descriptions of the view and had seen countless photos and movies of the Earth taken from space, I, like Glenn and most astronauts that followed, was unprepared for the beauty of it all. I was totally honored to have the opportunity to see the Earth as Glenn had first seen it decades earlier. The view was every bit as "tremendous" as Glenn had said.

Glenn's *Friendship 7* flight, along with follow-on missions involving Ohio astronauts Jim Lovell, Neil Armstrong, and others, was a tremendous influence for me and an entire generation of future space explorers. And in the pages that follow you will be able to experience some of the pride, drama, excitement, and thrills we all felt as we followed John Glenn's mission and began our dreams of flying in space ourselves one day.

Donald A. Thomas, Ph.D.
NASA Mission Specialist STS-65, STS-70, STS-83, and STS-94
Author of *Orbit of Discovery: The All-Ohio Space Shuttle Mission*

Acknowledgements

As I research each new book I always keep a running sheet listing the names of those who have assisted me in its compilation in order that I can thank them in the acknowledgements section. I am indebted to them for their kindness and willingness to help.

There are always the stalwarts; those who have helped me over many years with practical advice and suggestions, who supply me with information and photographs, and who read through my drafts in search of any obvious factual or grammatical errors. My continuing thanks therefore go to Joachim Becker (spacefacts.com), Kate Doolan, Francis French, Al Hallonquist, Ed Hengeveld, Robert Pearlman (collectSPACE.com), J.L. Pickering (Retro Space Images.com), and Eddie Pugh.

For responding to my requests for information, stories, or particular photographs, I am greatly indebted to Robert Bell, Erik Bergmann, the late Scott Carpenter, Dean Conger, Manfred ("Dutch") von Ehrenfried, Robert Frangenberg, Gene Kranz, Thomas Hanko, Donald Harter, Brian Harvey, Mike Humphrey, Jack Lousma, Lawrence ("Larry") McGlynn, Chris Miller, Teasel Muir-Harmony, Michael Neufeld and Roger Launius (both at the Smithsonian National Air and Space Museum), Dee O'Hara, Richard Pomfrey, Robert Truman, Charles Tynan, Dr. Stanley White, Milton Windler, and Eugene Wolfe.

Special thanks go to the NASA astronaut and accomplished author Don Thomas, a veteran of four space shuttle missions. Born and raised in the "Buckeye State," he was truly privileged to have his fellow Ohioan John Glenn pen a foreword to his sublime 2013 book, *Orbit of Discovery: The All-Ohio Space Shuttle Mission*, and he was delighted to repay the compliment by writing the foreword to this book.

I have now been associated with the good folks at Springer-Praxis for a number of years, and they are great (and fun) people to deal with. I therefore thank most sincerely Clive Horwood of Praxis Publishing in the United Kingdom; my superb copyeditor, fellow author, and good mate David M. Harland; and Jim Wilkie, who produced the glorious cover art for this and all my Springer books. At Springer Books, New York, my effusive and ongoing thanks to the hard-working Maury Solomon, Senior Editor, Physics and Astronomy, and her incredibly helpful Assistant Editor Nora Rawn, who has worked miracles for me in so many ways.

This book was written at a troublesome time for me, when I was (and still am) overcoming a serious double-knee injury as the result of a domestic accident which meant that I was not easily mobile for some time. So I must express my deep love and appreciation to Pat, my wonderful wife of 46 years, for not only helping me physically and supportively on the difficult path to recovery but for abiding me spending excessive amounts of time at my office keyboard when I would normally have been taking care of other, more pressing chores around the house.

I thank you, profoundly, one and all.

Author's prologue

It was Sunday, 15 July 1962; a date I remember well, and for all the right reasons. Back then, I was a 15-year-old Australian space enthusiast and NASA's Project Mercury was in full and glorious swing. I was following the space program with all the youthful zeal and devotion I could muster, and that momentous day in Sydney would dawn with a promise that, to me, was tantamount to a dream come true.

While it might have been a time of personal euphoria, in July 1962 global tension was rife. It was the height of the Cold War, the divisive Berlin Wall had recently been constructed, and it seemed that the Russians were well ahead in the Space Race. The world was just three months away from what would be the catastrophic prospect of global nuclear war resulting from the Cuban missile crisis. It was a time when the United States was in dire need of some good news and a patriotic boost, and Marine war hero John Glenn gave his nation both, as a timely and popular icon of renewed American pride.

Accompanied by two interested friends I had made my way to Queen's Square at the end of Sydney's Hyde Park, where later that day John Glenn's spacecraft *Friendship 7* would be on display over four days as part of a triumphant global tour of the historic vehicle. That morning a Military Air Transport Systems B-29 Globemaster touched down at Sydney's Kingsford Smith airport, having transported the capsule from an earlier public display in Perth – the city that turned on all its lights for a grateful orbiting Glenn as he soared over the Western Australian coastline by night just five months earlier.

Around lunchtime and amid much excitement, the history-making spacecraft arrived at Hyde Park, mounted on a special trailer, and was rapidly installed under a marquee with a short staircase leading up to and then down from the Plexiglas window, through which one could view the interior of the capsule. Inside, the astronaut's couch was occupied by a space-suited mannequin representing the astronaut. By 3:00 p.m. a line 450 yards long stretched across Hyde Park, but eventually it was our turn to mount the four steps. We were only permitted a few seconds each before being moved along by the uniformed attendants, but gazing inside that incredible piece of famed machinery provided me with one of the most memorable experiences of my life; certainly one of the best since I had first discovered a fascination with human space flight activity less than a year before. To me, John Glenn was the ultimate hero, a dynamic icon of our age, and a true inspiration.

The B-29 that brought the spacecraft to Australia was emblazoned with the words, "Around the world with *Friendship 7*' and the craft's so-called fourth orbit of the globe was depicted on a map of the four continents that the spacecraft would ultimately visit. (Photo: NASA)

Curious airport workers gather around as *Friendship 7* is unloaded from the U.S. Air Force cargo plane, ready for transportation into the city of Sydney. (Photo: Sydney Morning Herald)

After leaving Australia, *Friendship 7* went on display in Manila, Philippines, on 20 July 1962. The mannequin can clearly be seen inside the capsule. (Photo: NASA)

Space Shuttle *Discovery* roars into the Florida sky on 29 October 1998 carrying a crew of seven which included John Glenn, making his second flight into space. (Photo: NASA)

Following Glenn's Mercury flight, the *Friendship 7* spacecraft became nearly as famous as its pilot, traveling on its own celebratory global tour. In 17 countries around the world, including Australia, millions of people stood patiently in line to catch a glimpse of it. When the tour was over, the capsule took an honored permanent place at the Smithsonian Institution in Washington D.C., right alongside the Wright brothers' original history-making airplane and Charles Lindbergh's *Spirit of St. Louis*.

On 29 October 1998, some 36 years after viewing *Friendship 7* in Sydney, I was seated in the VIP stand at the Kennedy Space Center to witness the second launch of John Glenn, this time aboard Space Shuttle *Discovery* on mission STS-95. It was an event I was deter-mined not to miss, and I had flown all the way from Sydney the previous day through the kindness of a launch invitation sent to me by shuttle crew member Scott Parazynski, M.D, whom I had earlier met at a space school in Sydney. On that day my spirits soared along with *Discovery* as it lifted off the pad and blazed a brilliant trail of fire and smoke into the clear blue skies over Florida. I stood there in awe, exhilarated in the knowledge that my boyhood hero had once again ascended into the final frontier of space – and this time I was there to see it happen.

The story of John Glenn and his two flights into space has been told in meticulous detail in a veritable mountain of books, magazines, newspapers, and other publications, but his story and legacy still excite and inspire me, and many others. Today he is the last surviving member of the legendary seven Mercury astronauts, and it is a privilege for me to relate the story of his much-delayed and heart-stopping 1962 flight aboard the cramped spacecraft *Friendship 7*, in the sincere hope that it might serve to inspire a whole new generation of space enthusiasts.

Colin Burgess
Bangor, NSW, Australia

1

Developing the Mercury-Atlas program

The year was 1961. The National Aeronautics and Space Administration (NASA) had only been in existence for three years, but already two spectacularly successful space missions completed by Soviet cosmonauts were having a dramatic effect on the civilian space agency's carefully planned, step-by-step human space flight program.

A PROBLEM OF COMPLACENCY

The single-orbit flight of cosmonaut Yuri Gagarin in April 1961 hit America hard.

Unexpectedly divested of the historic prestige of placing the first person into space, NASA was accused of being too complacent. Every indication suggested that the Soviet Union was building up to this achievement, albeit with a program progressing under a strict shroud of secrecy.

This was in direct opposition to the open way in which NASA operated, with their events timetable openly announced. The Russians knew approximately when the first ballistic Mercury flight would take place, and even the names of the three prime astronaut candidates. With this and other technical knowledge available to their space chiefs, they were able to work to their own covert timetable and prepare to launch one of their cosmonaut team into orbit. They knew that the United States would only conduct a suborbital mission using a converted Redstone intermediate-range ballistic missile (IRBM), which was far less powerful than their own R-7 booster.

As a former administrator of NASA, James Edwin Webb later admitted in hindsight, the Russians had made several data-gathering, unmanned flights in early 1961. "We made a number of important, less spectacular, but very important flights. So both countries were coming down the line in flight programs that involved meteorological and other satellites.

"We were flying monkeys and they had earlier flown dogs. Soon thereafter they were flying men. But by and large there were a number of quite important flights that showed increased horizons as to what could be done. Now, it was perfectly clear that they had been flying a booster that could lift 10,000 pounds into orbit, and the biggest thing we could put up was a Mercury, which is about 3,000 pounds. It was perfectly clear that we were behind them, and that they had been working at least four or five years before we got started on these bigger boosters."[1]

© Springer International Publishing Switzerland 2015
C. Burgess, *Friendship 7*, Springer Praxis Books, DOI 10.1007/978-3-319-15654-5_1

President Kennedy shakes hands with NASA Administrator James E. Webb during a meeting in the Oval Office, 30 January 1961. (Photograph by Abbie Rowe in the John F. Kennedy Presidential Library and Museum, Boston)

BOOSTING "OLD RELIABLE"

In January 1959, just three months after NASA was established, the newly formed civilian space agency settled upon one of the most reliable large rockets ever produced in the United States, the Redstone, as the prime launch vehicle for its ambitious plans to place the first American astronauts – yet to be selected – into space. NASA also chose the McDonnell Aircraft Company to design and build the Mercury spacecraft.

A modified and enhanced version of Nazi Germany's deadly A-4/V-2 missile, the U.S. Army's Redstone was widely known as "Old Reliable" due to its renowned dependability and an impressive record of successfully completed launch and flight operations. Ahead of being selected by NASA, the Redstone had undergone several years of development and testing as a medium-range, tactical surface-to-surface missile for the Army Ballistic Missile Agency (ABMA) at what was called the Redstone Arsenal in Huntsville, Alabama. This reliability was a clear factor in its selection by the space agency.

With the cooperation of the Army, NASA issued a request to the ABMA for eight Redstone missiles to be launched in the first phase of what was now called Project Mercury, to one day send a number of astronauts on proving ballistic or suborbital space missions.

However, with a thrust of just 78,000 pounds, it was also recognized that the Redstone was incapable of inserting a manned spacecraft into Earth orbit. NASA was already looking beyond the ballistic program, and specifically at the Atlas series of rockets for their future manned orbital plans.

As history records, Navy Comdr. Alan Shepard was launched atop a Redstone rocket in his Mercury spacecraft *Freedom 7* on 5 May 1961. Although he became the first American to fly into space, he was beaten to the honor of the world's first spacefarer by cosmonaut Yuri Gagarin, blasted into a single Earth orbit in his Vostok spacecraft just three weeks earlier. Still, NASA was persisting in its plans for an incremental space flight program, with future manned Redstone launches in its provisional manifest.

Alan Shepard is launched on America's first manned space flight, 5 May 1961. (Photo: NASA)

Soviet cosmonauts Yuri Gagarin and Gherman Titov. (Photo: Author's collection)

With the fervor of an international space race rapidly gaining momentum, the United States and particularly President John F. Kennedy could no longer ignore events in the new frontier of space. On 21 May 1961, the President took up the political and patriotic challenge in a special message before Congress on "urgent national needs," in which he asked for an additional $7 billion to $9 billion over the next five years to upgrade the American space program.

"If we are to win the battle that is now going on around the world between freedom and tyranny," he declared, "the dramatic achievements in space which occurred in recent weeks should have made clear to us all, as did the *Sputnik* in 1957, the impact of this adventure on the minds of men everywhere, who are attempting to make a determination of which road they should take …. First, I believe that this nation should commit itself to achieving the goal, before this decade is out, of landing a man on the Moon and returning him safely to the Earth. No single space project in this period will be more impressive to mankind, or more important for the long-range exploration of space; and none will be so difficult or expensive to accomplish." With that, he had daringly thrown down the gauntlet to the Soviet Union.

But there were more shocks to follow. On 6 August 1961, following a successful Mercury suborbital space flight by U.S. Air Force Capt. Virgil ("Gus") Grissom – a flight in many ways similar to that of Shepard – NASA was once again left red-faced and self-questioning when the Soviet Union announced that cosmonaut Gherman Titov had been launched on an orbital space mission. His flight would last in excess of a day, and carry him through a total of 17 orbits of the planet. It was a massive propaganda coup for the Russians and their space efforts, and devastating news for NASA.

From left: John Glenn, Alan Shepard and Gus Grissom prior to Shepard's *Freedom 7* space flight. (Photo: NASA)

Under mounting pressure to speed things up, the space agency would re-evaluate its previous, carefully structured timetable. All of a sudden, "suborbital" became almost a dirty word in the NASA lexicon. The nation and its leaders were demanding that orbital missions begin immediately in an attempt to catch up to – and eventually surpass – the impudent but impressive efforts of the feared Russians and their largely unknown space technology.

UNVEILING ATLAS

In only its second week as a space agency, NASA had begun discussing possible applications of the Atlas rocket to the Mercury program. It was clearly recognized that the Atlas had not been designed as a human-carrying vehicle, but had been specifically developed for the U.S. Air Force as a military weapon capable of carrying nuclear warheads, and hence the launch vehicle would require substantial modifications.

Developing the Atlas rocket program had been a protracted, but top-priority task. It began in the late 1940s, when the Air Force appointed a company known as Consolidated Vultee Aircraft Corporation (more commonly called Convair) to produce a proposed long-range missile, the hardware part of Project MX-774. In this, the Air Force planned to create a rocket that would fulfill a demanding role as America's first intercontinental ballistic missile (ICBM).

Project MX-774 was shelved in 1947, although several vehicles were test-fired over the following two years by Convair, who had continued to research the test project with residual funds. In January 1951 the Air Force turned once again to Convair, awarding the company a substantial contract to produce a rocket-powered ballistic missile for Project MX-1593.

The single-stage test-bed prototype produced by Convair (which became the Convair division of General Dynamics in 1954) was initially known as the XSM-16A, but later redesignated the X-11 (Atlas). The contract called for the manufacture of twelve missiles; three of which would be used purely for captive or static test firings.

The liquid-fuelled Atlas was manufactured at Convair's Kearny Mesa plant in San Diego. Powered by rocket-grade RP-1 (highly refined kerosene) with a liquid oxygen oxidizer, it proved to be a vehicle of amazing contradictions. Weighing around 267,000 pounds when fully fuelled, the key feature of the missile's design was its extremely lightweight structure, conceived and designed for the most part by the imaginative Belgium-born Convair engineer, Karel J. ("Charlie") Bossart. The stainless steel skin of the Atlas was actually thinner in parts than a modern compact disc; so thin in fact that without the use of pneumatics to keep it erect, an unfueled Atlas would have rapidly collapsed under its own weight. The rocket would retain its shape when the highly pressurized tanks were filled with rocket fuel, but when empty, the tanks maintained their shape through pressurization from helium or nitrogen gas pressure; there was no stiffening by internal framework. The tanks had to be filled with gas at a positive internal pressure of 5 psi in order to maintain both rigidity and integrity.

Atlas rockets under construction at the Convair/General Dynamics Kearny Mesa plant. (Photo: San Diego Air & Space Museum)

Cutting the thin metal skin for an Atlas rocket. (Photo: San Diego Air & Space Museum)

German-born rocket designer Wernher von Braun, then at the Redstone Arsenal in Huntsville, Alabama, actually felt that the Atlas rocket would not survive the stresses of launch, derisively calling it no more than a blimp, but fortunately he would be proven wrong.

The formal production of the Atlas commenced in January 1955, with captive test firings carried out at Edwards Air Force Base, California, the following year. A twelve-month flight test of eight single-engine prototypes then took place at Cape Canaveral, beginning in June 1957. The first Atlas-A launch on 11 June (Atlas 4A) ended in spectacular failure, owing to a problem in the booster system. The errant missile was remotely destroyed by the range safety officer, who would later press the red button to detonate explosives aboard a second wayward Atlas-A launched in September.

A successful launch and ballistic trajectory finally took place on 17 December, coinciding with the fifty-fourth anniversary of the Wright brothers' first successful flight. The following year, during a mixed launch program of successes and failures, an entire Atlas – now produced with three engines – reached orbit carrying the United States' first communications satellite, SCORE (Signal Communications by Orbiting Relay Equipment). On 19 December 1958 communications history was created when the satellite beamed down to Earth a pre-recorded 30-second Christmas message from President Dwight Eisenhower.

Atlas 6A on Launch Complex LC-14 prior to the 25 September 1957 launch. (Photo: NASA)

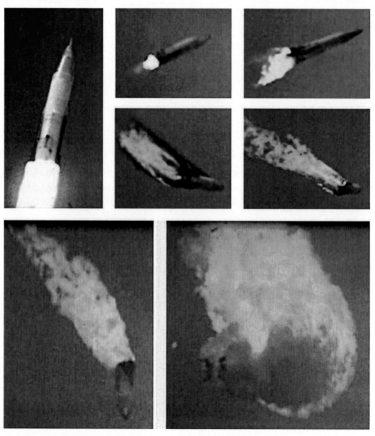

When Atlas 6A became erratic after lift-off it had to be destroyed by the range safety officer.
(Photos: NASA)

POWERING TO ORBIT

An awe-inspiring amount of power was needed to hoist a fully fuelled Atlas missile off the launch pad. The now-upgraded, 82-foot ICBM, officially deployed in September 1959, was powered by two large booster engines that burned RP-1 and liquid oxygen, producing a combined 360,000 pounds of thrust. These flanked a smaller, single sustainer engine which developed a further 57,000 pounds of thrust. Additionally, there were two small steerable vernier engines mounted above the sustainer engine.

All five engines were explosively ignited at the moment of lift-off, but only one of the three main engines would remain with the Atlas all the way to orbit. The two outer booster engines would only fire during the first 140 seconds of flight, after which they cut out and were jettisoned. After shedding this weight the Atlas was now light enough to achieve orbit. Meanwhile the sustainer engine continued to burn for a further 130 seconds, accelerating the rocket until it attained a velocity of about 16,000 miles per hour. It then shut down and the small vernier rockets trimmed the velocity to the exact value required, and made orientation corrections until they too were shut down.

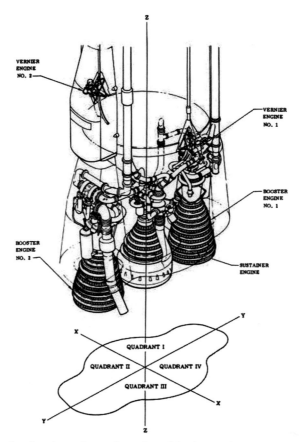

This schematic drawing shows the configuration of the five engines used on the Atlas rocket. (Photo: NASA)

The Mercury program reached an important milestone on 9 September 1959 with the successful night-time launch from Cape Canaveral of what was called "Big Joe," carrying a full-size instrumented "boilerplate" mockup of the proposed Mercury spacecraft. The launch vehicle, designated Atlas 10D, was also the first Atlas rocket to be used in a Project Mercury launch. The flight was designed to test the aerodynamic stability of the capsule as well as the effectiveness of its ablative heat shield. Although lift-off from Launch Complex LC-14 went well, the two outer booster engines failed to jettison owing to a malfunction, and the Big Joe/Atlas assembly only reached an altitude of just under 100 miles. Nevertheless this was sufficiently high to test the heat shield, and once the capsule had separated from the spent Atlas booster it fell back to Earth in conditions simulating a return from orbit. On this proving flight there was no retro-package attached to the heat shield as there would be on a manned mission. After splashing down, the Big Joe capsule was retrieved off the coast of Puerto Rico by the crew of the Navy destroyer, USS *Strong* (DD-758). Data later confirmed that the blunt-body capsule shape and ablative heat shield had performed as predicted, and the Big Joe test flight was a major step in proving that the Mercury design concepts were fully capable of the task ahead. A second test flight, Big Joe II, was deemed unnecessary and the Atlas rocket was deployed elsewhere.[2]

Project Mercury Astronaut Training Officer and Navy psychologist Robert Voas has many fond recollections of that critical period in the impetus to launch a manned spacecraft.

"One of my special memories of that early period would have been probably in the fall of 1959. It was the first time we took the astronauts to the Cape, and we watched from quite nearby a launch. The wonderful thing about the Cape … in those days when you were more intimately related to the launches in some ways than now, they're so large and involved, so big that I think you're more separated from them. But whoever started that whole process had guidance from Cecil B. DeMille and the movie colony, you know, because they would launch early in the morning, but all night they would be doing the fueling. So you had this large silver rocket with spotlights on it, and the LOX [liquid oxygen] is boiling off, so steam is coming off of it, and it's just a spectacular sight.

"Then as the launch comes, why, at first they tie it down so that the engines are fully operational before they release. But you see this great fireball come up. It's somewhat reminiscent of the pictures we used to see of the atomic burst. Then out of it comes the silver rocket. So it's a very impressive thing, and I remember that first time when the astronauts –you could see their faces thinking about they were going to be sitting on top of that, riding it away. But that was quite a fun experience."[3]

The Mercury-Atlas 10D Big Joe rocket sits on Cape Canaveral's Launch Complex LC-14, ready for lift-off on 9 September 1959. (Photo: NASA)

Despite some spectacular and successful launches the Atlas rocket – once described as the most complex operational machine ever invented by man – initially earned a reputation as a notoriously volatile vehicle. By the time John Glenn had been selected to make America's first manned orbital flight, one out of every three Atlas launches was ending in catastrophic failure, including the one that would precede his mission. This was despite NASA setting the Mercury-Atlas reliability standard so high that few industry observers believed it could be achieved. But the space agency set out to make it a reality. Together, NASA and General Dynamics Astronautics (formerly Convair-Astronautics) instituted the most relentlessly thorough reliability procedures ever applied to a vehicle. The Atlas would later achieve an agreeable reputation as the formidable steed that safely propelled four of NASA's Mercury astronauts into orbit.

The spectacular launch of Atlas 14D on 11 August 1959 from Launch Complex LC-13 at Cape Canaveral. (Photo: NASA)

On 13 May 1961 an advanced model Atlas flew 5,000 miles on its second successful flight. The series "E" Atlas, boosted by 398,000 pounds of thrust, lifted off from Launch Complex 11 at Cape Canaveral Air Force Station on what was later called a "completely successful" test of its engines and guidance system. The Atlas 12E carried a data package released from its nose cone upon re-entering the atmosphere, later recovered at sea. In San Diego, Karel Bossart, technical director of General Dynamics Astronautics, said the missile had now proved itself and was set for the next major step. "We are ready to deliver an astronaut into orbit whenever NASA is ready," he stated with confidence.[4]

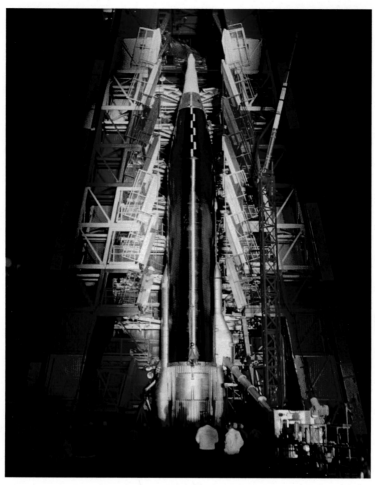

Atlas 12E surrounded by the gantry on Launch Complex LC-11, 13 May 1961. (Photo: NASA)

The modified launch vehicle used on John Glenn's MA-6 mission was an Atlas D model. There were several principal differences between the military version and the one used in the Mercury-Atlas program. The payload adapter section had been modified to accommodate the Mercury spacecraft, the structure of the upper neck of the Atlas had been strengthened to deal with the increased aerodynamic stresses, and an automatic Abort Sensing and Implementation System (ASIS) had been installed to detect any deviation in the performance of the Atlas and trigger the Mercury Escape System before an impending catastrophic failure.[5]

Karel ("Charlie") Bossart of General Dynamics Astronautics, considered to be the "Father of the Atlas." (Photo: San Diego Air & Space Museum)

THE LITTLE JOE TESTS

One of the important issues for NASA during the development of the Mercury-Atlas program was the need to know that an astronaut in peril could be saved by promptly removing him from the danger area in the event of a catastrophic failure on the launch pad or during early ascent. This would require verification tests of the Mercury spacecraft's escape system.

The escape system that had been devised was somewhat rudimentary, but extremely effective. A small but powerful rocket with three exhaust bells, canted out at precise angles, was attached to a red 16-foot tower mounted above the Mercury spacecraft. The tower could be fired during a manned launch if a potentially catastrophic situation evolved, hauling the capsule free of a malfunctioning booster. Testing the complex sequence of events associated with the escape system required the use of a suitable, uncomplicated launch test vehicle.

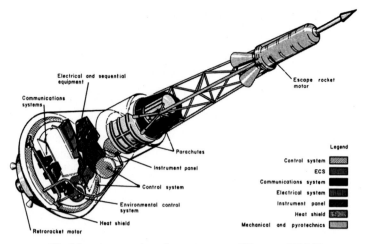

The Mercury capsule and escape tower. (Diagram: NASA)

A Lockheed engineer works on the assembly of the Mercury escape tower. (Photo: Lockheed Corporation/NASA)

Prior to NASA being established in 1958, and with the realization that human-tended space flight would soon become a reality, engineers and scientists at the Langley Research Center in Virginia developed a solid-propellant launch vehicle for testing prototype Mercury capsules and the associated rocket escape system. According to a contemporary NASA fact sheet, the booster system "had to be simple in concept, use existing proven equipment wherever possible, be flexible so that it could perform a variety of missions, avoid the use of complex systems, be as inexpensive as possible, and be designed in such a manner as to keep ground support requirements at a minimum."[6] Thus the "Little Joe" booster would emerge from Langley, its series nickname being inspired by its stubby appearance in comparison to other rockets from that period.

Little Joe rockets were capable of creating conditions during the initial flight through the atmosphere that would closely approximate those of the much larger and more power-ful Atlas, which was destined to become the launch vehicle used during Project Mercury's orbital flight missions.

In March 1959, U.S. Air Force Headquarters directed the School of Aviation Medicine (SAM) in San Antonio, Texas to provide biomedical support for Project Mercury launches at the invitation of NASA's Space Task Group (STG). The STG needed to conduct tests of their equipment and procedures relating to an emergency-induced separation of the launch vehicle and the Mercury spacecraft.

Originally, a task group attached to the Air Force had planned to conduct two proving flights of the proposed Mercury capsule using the reliable Jupiter booster, with the second flight carrying a primate passenger to qualify the environmental control system (ECS). On 1 July 1959, according to a report on Mercury primates by then Executive Director of the Alamogordo Space Center, Gregory P. Kennedy: "Abe Silverstein, Director of Space Flight Development at NASA Headquarters, sent a formal memo to the Space Task Group cancelling the Mercury Jupiter program." However, he added that "the concept of testing the spacecraft with primates remained, and during the same time frame that the Mercury Jupiter was being terminated, the animal test program took shape."[7] The 6571st laboratory at Holloman Air Force Base in New Mexico was training young captured primates for use in aerospace experimentation, and the facility was subsequently directed to provide suit-able animals for up to four planned flights.

Earlier, on 26 May, SAM was principally tasked with manufacturing, fitting out and testing several airtight capsules that could safely restrain and offer life support to a primate within a full-size mockup of the Mercury capsule. During the flight, designated Little Joe-2 (LJ-2), the primate passenger would wear a complex space suit that had been devel-oped at the school to measure the animal's physiological reactions and conditions while ensconced within a fiberglass biological capsule, or biopack. Due to several factors includ-ing capsule space and weight constraints, technicians and scientists determined that it would be far more practical to train and fly six- to eight-pound Rhesus monkeys, rather than the larger chimpanzees.[8]

While the geometric configuration of the capsule would duplicate the configuration of a production Mercury craft, the structural materials were quite different. This much sim-plified model, manufactured at Langley, became known as a "boilerplate" version.

The program would undertake two ballistic flights. The first was to test the escape system at high altitude, while the second would simulate an emergency separation when dynamic pressures or air loads were at their maximum during ascent.

The Rhesus monkey selected as the most suitable candidate was named Sam, after the acronym of the School of Aviation Medicine. On the afternoon before launch day, his body plastered with monitoring sensors, Sam was carefully strapped onto the biopack couch, then he was given a pre-flight snack of an apple and some orange juice.

The next morning, two interested spectators were on hand for the LJ-2 launch; Mercury astronauts Alan Shepard and Virgil ("Gus") Grissom. Shepard's astronaut duties involved a study of launch escape systems, so he was keen to witness how well (or otherwise) the equipment and procedures worked.

Sam is inserted into his biopack container prior to the Little Joe-2 launch. (Photo: NASA)

SAM AND MISS SAM

At 11:15 a.m. on 4 December 1959, the Little Joe carrying Sam was launched from the pad at the Wallops Island Flight Test Range in Virginia. At 59 seconds into the flight, with the booster's fuel fully expended, an abort sequence was initiated by timers.

The motors on the escape tower fired for just one second but with incredible power, and pulled the capsule away from the booster rocket. The capsule/escape tower assembly then continued to coast upwards to an apogee of 53 miles – about 15 miles less than hoped for – before the tower and rocket-motor case were jettisoned on cue from the capsule. Gravity then began acting to drag the spacecraft back down to the ground in a mild re-entry. The drogue and main parachute systems operated perfectly at just over 20,000 feet and 10,000 feet respectively, and the capsule splashed down hard but safely into the Atlantic Ocean off Cape Hatteras, North Carolina.

Two hours after the Little Joe booster had left the launch pad, crewmembers on board the recovery ship USS *Borie* (DD-215) hooked onto the capsule and hoisted it aboard. When Sam was removed from the biopack sometime later, he was found to be in fine shape.

The launch of Little Joe-2 from Wallops Island. (Photo: NASA)

USS *Borie* crewmembers retrieve the LJ-2 Mercury capsule. (Photo: NASA)

While he had experienced only three minutes of weightlessness, more importantly both Sam and the Mercury capsule's escape systems had come through the exercise with flying colors. Shepard and Grissom were understandably delighted.

Another primate test flight loaded with a simulated ascent "accident" was scheduled to take to the Virginia skies six weeks after the successful conclusion to Sam's mission. Prior to this, however, another Little Joe flight occurred with less than satisfactory results. This fully instrumented repeat flight without a monkey was designated LJ-1. It was scheduled to lift off on 21 August to test a launch abort under high aerodynamic load conditions. Just 31 minutes before launch the capsule's escape tower suddenly fired due to a faulty escape circuit. As the booster remained on the launch pad, the escape tower dragged the boiler-plate capsule to an altitude of 2,000 feet, sending ground crews and photographers scurrying for cover. Fortunately no one was injured.

Another Little Joe proving flight (LJ-6) designed to test the spacecraft's aerodynamics and integrity was carried out on 4 October. On this occasion the non-instrumented boilerplate capsule was fitted with an inert escape rocket system. After two and a half minutes of flight the rocket was intentionally destroyed to prove the effectiveness of the destruct system.[9]

The Little Joe-1 launch vehicle at Wallops Island topped by a Mercury capsule, August 1959.
(Photo: NASA)

With repairs and systems overhauls effected, the earlier unsuccessful Little Joe/
Mercury LJ-1 launch was redesignated LJ-1A and took to the skies on 4 November. The
purpose of this second attempt was to test the escape system in conditions equivalent to
the maximum dynamic pressure of an Atlas flight, or about 1,000 pounds per square foot.
The spacecraft contained a pressure-sensing device that should have initiated a planned
abort thirty seconds after launch, but when this system fired the escape motor igniter, it
took a few seconds for sufficient thrust to build up and the abort sequence was not accomplished at the desired dynamic pressure. The mission was written off as only a partial
success.

Miss Sam prior to her Little Joe-1B launch. (Photo: NASA)

The fourth Mercury/Little Joe test flight (LJ-1B), which was planned to overcome the abort system problems encountered on LJ-1A, would carry a Rhesus monkey named Miss Sam. There were five main objectives manifested for this flight:

1. To check the Mercury escape system concept and hardware at the maximum dynamic pressure anticipated during a Mercury-Atlas exit flight [i.e. ascent].
2. To determine the effects of simulated Atlas abort accelerations on a small primate.
3. To obtain further reliability data on the Mercury spacecraft drogue and main parachute operations.
4. To check out the operational effectiveness of spacecraft recovery by helicopter.
5. To recover the escape-system assembly (escape motor and tower) for a post-flight examination to determine if there were any component malfunctions or structural failures.[10]

Launch of LJ-1B with Miss Sam occupying the boilerplate capsule. (Photo: NASA)

On 21 January 1960 Miss Sam was rocketed to a nominal altitude of just over nine miles from the Wallops Island Flight Test Range, the Little Joe booster achieving a maximum velocity of just over 2,000 miles per hour. The extreme force of the launch pressed her back into the support couch at nearly fourteen times her usual body weight, but she seemed to endure this with minimal physiological disturbance.

When the escape sequence was initiated, to haul the boilerplate capsule free of its booster, Miss Sam was thrown about by the unexpected thrust of the escape rocket. She also had to contend with a higher than anticipated noise level in the capsule.

Eight and a half minutes after Little Joe-1B lifted off from Wallops Island, during which time the escape sequence system and the landing systems functioned perfectly, the boilerplate spacecraft splashed down smoothly 12 miles from the launch site. This time, instead of being picked up by a recovery ship, a waiting Marine Corps helicopter plucked the spacecraft from the Atlantic and carried it back to Wallops Island. A medical examination soon afterwards indicated that Miss Sam was in excellent health.

While only two of the six eventual Little Joe qualification test flights would carry primate passengers, the success of the Little Joe-1B test flight meant that the next launch in the series (LJ-5) would be the first to fly an actual spacecraft off the production line in the McDonnell plant in St. Louis, Missouri.[11]

JOHN GLENN NAMED TO MA-6

The fourth anniversary of the launch of *Sputnik*, 4 October 1961, proved to be momentous for NASA astronaut John Glenn, as he later reflected.

"Naturally, it was a great moment for me when the announcement came that I would be the first American to orbit the Earth. I first knew it when Robert Gilruth, the Director of

Project Mercury, called the seven of us into his office at Langley and told us that I would be the pilot on MA-6 and that Scott Carpenter would be the backup pilot. After that we all stood up, and one by one the other men came over and shook my hand. We are not a bunch of back-slappers. You are always happy for the guy they pick and sorry if it's not you. Then you congratulate the pilot, and that's all there is to it."[12]

Months earlier, Glenn had been forced to swallow his extreme disappointment and put on a brave face in public when Alan Shepard was selected to make the first U.S. suborbital space flight, basically a 15-minute ballistic proving shot, with Gus Grissom assigned to the second Mercury mission, MR-4. Although he was chosen to back up both missions and was slated to make the third suborbital Mercury-Redstone mission, Glenn considered it as something of an inexplicable rejection of his abilities and track record as a test pilot.

"I guess I am a fairly dogged competitor," he admitted, "and getting left behind twice in a row was a little like always being a bridesmaid but never a bride."[13]

Nevertheless, Glenn threw himself into his backup role with determination, and certainly benefitted from this experience, as well as from what Shepard and Grissom learned on their suborbital missions. Following the orbital flights of cosmonauts Gagarin and Titov, NASA canceled the remaining suborbital flights and began preparing for an orbital mission, and it was John Glenn who was appointed to the first flight.

"From a technical standpoint, the orbital mission would be quite different from the ballistic flights in several respects. For one thing, we would be using the Atlas missile as a launch vehicle in order to get up the required thrust and velocity to get into orbit. The Atlas' engines have a total thrust of 360,000 pounds compared to 76,000 for the Redstone, and the Atlas would get the capsule up to a top speed of nearly 18,000 miles per hour, which is more than three times the speed of the Redstone IRBM, which had served us well on the ballistic flights. Once in orbit, the flight would also last longer – about four and a half hours if we made all three orbits. This would be a new magnitude of space flight for the U.S., and if it were successful it would pave the way for longer voyages, eventually to

Gus Grissom, John Glenn and Alan Shepard in their Mercury spacesuits. (Photo: NASA)

the Moon and beyond, just as Al and Gus had paved the way for this one. It would be a prelude to our plans for the future."[14]

When asked, Scott Carpenter was equally philosophical. "There was really no difficulty between the suborbital guys and the orbital guys. Everybody wanted to take the next flight, whatever it was, and so I remember thinking at the time that we were all in a heated competition with each other, but the Three Musketeers came to my mind – we were all for one and one for all. And it worked without failure where helping to get the program under way and going steadily was concerned. We had petty difficulties here and there about other things, but we were all one team where the program was concerned."[15]

The Mercury capsule in which Glenn was to make his historic journey into space was already at the Cape. It arrived towards the end of August, ready to be prepared and used in training for the first orbital flight, and the mission calendar at that time called for a launch in December. Little did anyone know that a number of frustrating delays would ensure that the much-anticipated mission would not fly until February the following year.

MA-6 spacecraft *Friendship 7* at Cape Canaveral. (Photo: NASA)

1961 finally rolled over into 1962, with much to anticipate on the NASA calendar. On 11 January, in giving his State of the Union message to the Congress, President Kennedy said:

"With the approval of this Congress, we have undertaken in the past year a great new effort in outer space. Our aim is not simply to be first on the Moon, any more than Charles Lindbergh's real aim was to be first to Paris. His aim was to develop the techniques and the authority of this country and other countries in the field of the air and the atmosphere.

"And our objective in making this effort, which we hope will place one of our citizens on the Moon, is to develop in a new frontier of science, commerce and cooperation, the position of the United States and the free world. This nation belongs among the first to explore it. And among the first, if not the first, we shall be."[16]

REFERENCES

1. T.H. Baker, "An Interview with James Webb: Administration of Exploration," from *Quest: The History of Spaceflight Quarterly* magazine, issue Vol. 21, No. 3, 2014, pp. 21–36
2. James R. Hansen, *NASA Spaceflight Revolution: NASA Langley Research Center From Sputnik to Apollo*, NASA History Series SP-4308, NASA Headquarters, Washington, D.C., 1995
3. Robert B. Voas interview with Summer Chick Bergen for Johnson Space Center Oral History Project, Vienna, Virginia, 19 May 2002
4. *Tonawanda News* (New York), article "U.S. scores 2nd success with Atlas," issue 13 May 1961, pg. 1
5. NASA *Space News Roundup* article, "Launch Vehicle for MA-6 Test is described," Manned Spacecraft Center, Houston, TX, issue 7 February 1962, pg. 2
6. Undated NASA Fact Sheet, *Project Mercury Little Joe Test Program*, issued by NASA Headquarters, Washington, D.C.
7. George P. Kennedy, *Mercury Primates* (IAA-89-741), The Space Center, Alamogordo, NM, 1989
8. Paul Laser, "Oral History of Brooks Air Force Base," HHM Inc., Austin, TX. History segment *Man-in-Space Program*. Website: *http://www.brooks.af.mil/history/space.html*
9. Hal T. Baber, Jr., Howard S. Carter and Roland D. English, NASA Technical Memorandum: *Flight Test of a Little Joe Boosted Full-Scale Spacecraft Model and Escape System for Project Mercury*, NASA TM X-629. Declassified 28 June 1967
10. Joseph Adams Shortal, NASA Reference Publication, *A New Dimension: Wallops Island Flight Test Range, the First Fifteen Years*, December 1978
11. *Ibid*
12. Carpenter, S., Cooper, Jr. L, Glenn, Jr., J., Grissom, V., Schirra, Jr., W., Shepard, Jr., A., and Slayton, D., *We Seven*, Simon and Schuster Inc., New York, NY, 1962, pp. 303–305
13. *Ibid*
14. *Ibid*
15. Scott Carpenter telephone interview with Colin Burgess, 18 December 2002
16. Public Papers of the Presidents of the United States: John F. Kennedy, 1962

2

The precursory flight of chimpanzee Enos

Much to the chagrin and bewilderment of Alan Shepard, a trained chimpanzee named Ham was launched on a proving suborbital flight aboard a Mercury capsule before the Navy astronaut was given a similar chance to fly into space. He felt it was an unnecessary step. Once, when asked why he had been selected as America's first astronaut, he wryly quipped, "I guess they ran out of monkeys!"

HAM'S MERCURY MISSION

On 31 January 1961, Ham was launched atop a Redstone booster on the suborbital Mercury-Redstone MR-2 flight. As John Glenn later noted, even though Ham's flight did not exactly go according to plan, the engineers learned some valuable lessons from the flight and were able to make vital changes to the Mercury spacecraft's systems ahead of a manned orbital flight.

"The scientists had decided to use a chimp as a stand-in because this particular animal closely resembles man from a physiological standpoint and can be trained to take part in a scientific experiment," Glenn observed. "A chimp's reaction to various stimuli, incidentally, is seven-tenths of a second, which is fairly close to the average man's reaction time of five-tenths of a second. Ham had been taught to perform a few simple tricks in order to test his ability to carry out useful functions during the four and a half minutes of weightlessness that he would be subjected to."

During the ballistic flight Ham would watch a series of flashing lights and pull various levers in sequence when the lights came on. A supply of water and banana-flavored pellets was installed inside his chamber to reward him for his prowess on the levers and keep him happy.

"He was strapped into a small couch which resembled the contour couch we would use," Glenn added, "and he was encased in a pressurized plastic chamber about the size of a small trunk which was fastened in place inside the capsule and connected with the oxygen supply. The chamber was sealed off from the cabin of the capsule, just as a man would be sealed off inside his pressure suit. Ham also had medical sensors attached to his body, much as we would, to record his heart rate, respiration and body temperature during the flight."[1]

© Springer International Publishing Switzerland 2015
C. Burgess, *Friendship 7*, Springer Praxis Books, DOI 10.1007/978-3-319-15654-5_2

Ham had to endure a troubled mission. At lift-off, a faulty valve in the booster led to the fuel pump injecting far too much liquid oxygen into the engine, causing the rocket to over-thrust and accelerate faster than expected. The fuel was rapidly depleted, which triggered an abort. As a result, the escape tower mounted above the capsule fired, ripping the capsule away from the spent booster. Ham was immediately subjected to crushing forces of around 17 g's. The chimpanzee managed to survive all these dramas and was eventually recovered at sea, but when he was extracted from the capsule Ham demonstrated to his rescuers that he was far from happy with the whole situation.

When Shepard later reviewed telemetry tapes and other data from MR-2 he was unshak-ably confident he could also have survived the flight. As plans then stood, he was due and ready to be launched on his own Mercury-Redstone flight (MR-3) on 24 March. But Wernher von Braun at the Redstone Arsenal in Huntsville, Alabama, and others of influ-ence were deeply concerned about the erratic performance of the Redstone booster on the MR-2 flight. After modifications had been completed they called for a further, unmanned test. The German rocketeer got his wish; the MR-BD (Mercury-Redstone Booster Development) flight was inserted into the schedule and was an entirely successful opera-tion. Shepard's flight was next in line. However, just three weeks ahead of his MR-3 flight, Soviet cosmonaut Yuri Gagarin was launched into orbit, becoming the first person to fly into space. Like Ham before him, but for entirely different reasons, Shepard was under-standably irate.

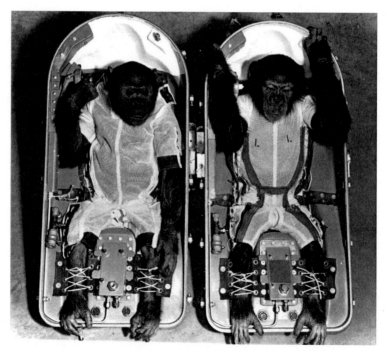

The two Holloman-trained chimpanzees that would complete Mercury flights: Enos (left) and Ham. (Photo: NASA)

Despite all the dramas associated with Ham's flight, the 3-year-old chimpanzee had convincingly shown that a human being could similarly survive the dynamics of rocket launch and re-entry, and function quite normally in a weightless environment.

Both Alan Shepard and Gus Grissom (MR-4) flew suborbital missions that meant they were only weightless for about five minutes, allowing them very little time to explore this phenomenon. With virtually no gravity at orbital height, NASA was anxious to ensure that an astronaut could perform simple tasks while weightless for an extended period. Some scientists were still concerned that orbiting astronauts, on seeing the planet passing so quickly beneath them, might even lose their normal concepts of up and down, speed and direction, and become critically disoriented. Once again, it was decided to conduct a precursory flight – this time orbital – using a suitable chimpanzee from the animals undergoing training in the special facility administered by the 6571st Aeromedical Research Laboratory at Holloman Air Force Base in New Mexico.

TRAINING THE CHIMPS

Whereas the Redstone chimpanzee candidates had been tested for their ability to remember commands during a ballistic flight through the use of a training device called a psychomotor panel, which dispensed a banana pellet if the animal pressed the correct lever when given a lighted cue, the unit used for training an orbiting animal was rather more complex. This advanced psychomotor involved the use of colors and symbols. In training the candidate chimpanzees, technicians could illuminate three symbols – circles, triangles, and squares. When two symbols having the same shape lit up on a panel facing the animal, it had to press a button below the non-matching third symbol. As a reward, every time the chimpanzee got it right a banana-flavored pellet popped out of a tube near its mouth. "They had to hit the lever so many times for a drink of water and so many times for a banana pellet," former aeromedical technician Master Sergeant Ed Dittmer pointed out.[2] At other times a small green light would illuminate during the tests. If the subject chimp noticed this and pressed a button to turn it off within 20 seconds, their reward was a drink of water or fruit juice from another tube situated near their mouth. But if any task was overlooked there was a penalty to pay; a small but unpleasant electrical shock would tingle a plate attached to the chimp's feet to demonstrate that something had not been done correctly.

The primate candidates also learned how to turn display lights on and off by hitting right- and left-hand levers, and were taught to count by pulling another small lever exactly fifty times. As they grew conditioned to the test sequence the animals would reach their own count of forty-nine, deliberately slowing as they approached the end of their task, and then eagerly place a hand under the tube for their banana pellet reward on the last pull. Trainers were amazed at how quickly their playful charges learned this routine and how they hardly ever got it wrong.

It was important for the upcoming MA-5 flight that the chimpanzees memorized the order in which they were to perform such tests, as one chimp would be replicating them in orbit. It followed that if an extended space flight had no effect on the ability of a primate

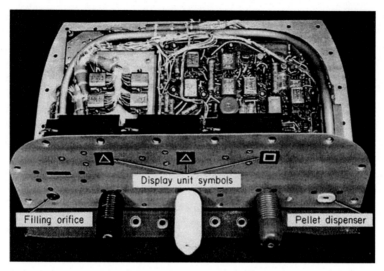

A psychomotor similar to the one installed on the MA-5 flight. (Photo: NASA)

to conduct simple sequential tasks, then there was no doubting a human's ability to do the same.[3] Yuri Gagarin seemed – at least back then – to have completed his orbital mission without any physiological problems, but garnering any reliable contemporary data on his flight was a near impossible task.

A SUITABLE CANDIDATE FOR ORBIT

On 13 September 1961, ahead of the chimpanzee flight, an unmanned test mission began with the launch of MA-4 from Cape Canaveral. A mechanical "simulated astronaut" was wired into the spacecraft as part of an extensive systems check, consuming oxygen and expending carbon dioxide at the same rate as an astronaut. Having completed a planned single orbit of the Earth the capsule was successfully recovered after splashdown.

Things were moving along. On 29 October 1961, the 6571st Aeromedical Research Laboratory delivered three of their best-trained chimpanzees to Cape Canaveral as potential candidates for MA-5, accompanied by twelve handlers. They linked up with eight handlers and another two chimps already in training at the Cape. In addition to space veteran Ham, the animals involved were Duane and Jim (named after project veterinarians Duane Mitch and James Cook), Rocky (named for boxer Rocky Graziano), and Enos. It soon became apparent that Ham, having endured a drama-filled ballistic flight, was less in contention for the role this time. Although he undertook his tests with a disappointing lack of enthusiasm, Ham nevertheless remained one of the three prime candidates just two days before the flight. However the 39-pound Enos (previously known by the designation No. 81) had proved to be a superior and intelligent candidate, although his handlers found him to be something of a handful at times.

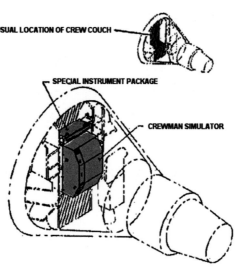

USUAL LOCATION OF CREW COUCH

SPECIAL INSTRUMENT PACKAGE

CREWMAN SIMULATOR

A schematic of the MA-4 capsule, showing the location of the crewman simulator. (Illustration: NASA)

Launch of the unmanned MA-4 proving flight. (Photo: NASA)

A wary handler with the unpredictable Enos. (Photo: NASA)

Enos (which means "man" in Hebrew) was a feisty, 4-year-old *Pan satyrus* chimpanzee. Although he was the star pupil, Enos often misbehaved badly and was far less tolerant of his human handlers than Ham. "No one ever held Enos," Ed Dittmer recalled. "If you had him, he was on a little strap. Enos was a good chimp and he was smart, but he didn't take to people. They had the wrong impression of him; they said he was a mean chimp and so forth, but he just didn't take to cuddling. That's why in any pictures you ever see of Enos you don't see anyone holding him."[4]

Enos became notorious for dropping his training diapers and stroking his genitalia whenever reporters paid a visit to the sheet-metal building where the chimpanzees were housed. It was this unsavory practice that caused him to be given the nickname of "Enos the Penis." The primate enclosure was right next to Hangar S, where the Mercury astronauts worked and trained, and where their offices were located. To minimize any disruption for the animals, visitors were discouraged, but every so often someone would want to see the chimp colony.

On one occasion a visiting politician somehow managed to pull a few VIP strings and was given a tour of the facility by a reluctant McDonnell launch pad leader, Guenter Wendt. "He wanted to see the monkeys," Wendt recalled. "I told them they weren't monkeys, they were highly intelligent chimpanzees. But he insisted on seeing the monkeys." Wendt, knowing all too well Enos's notoriously bad temperament, attempted to talk the congressman out of the viewing, especially as the chimpanzee had only just come back from a training session and was in a particularly foul mood.

"When those chimps didn't do what they were supposed to do, they'd receive a little shock through their feet," Wendt said. "So I knew this wasn't a good time to see Enos. But this guy said, 'I'm a congressman, and I want to see the monkeys.' So I took him, even though I knew exactly what was going to happen." Wendt dutifully opened the door leading into the chimp's quarters and ushered the congressman in first. Enos, upon seeing a stranger enter his abode, immediately squatted down, emptied his bowels into his hand, and flung a pile of steaming feces at the congressman, who recoiled in horror, his pristine white shirt and suit splattered with the stinking mess. He later sheepishly admitted to Wendt, "I can see why you didn't want me to see him." Wendt doesn't believe the congressman was ever a guest again at the Cape.[5]

Eventually, despite his wayward behavior, it was decided that Enos would fly on the three-orbit MA-5 mission.

Enos relaxing in his form-fitting flight couch. (Photo: NASA)

ENOS AND MA-5

Originally scheduled for lift-off on 7 November 1961, the MA-5 flight was delayed a week before being indefinitely postponed on 11 November as a result of a hydrogen peroxide leak in the spacecraft's manual control system. By this time NASA's future planning had been further thwarted following Soviet cosmonaut Gherman Titov's day-long mission four months earlier. As a consequence, there had been urgent calls from within and outside NASA for the MA-5 flight to be shelved and for John Glenn to be launched on America's first orbital flight before the end of the year. However good sense prevailed; it was crucial that additional flight data be gathered and for systems tests to be carried out before an astronaut could be launched on top of the unpredictable Atlas rocket.

Enos's MA-5 Mercury capsule being mated to the Atlas booster. (Photo: NASA)

When asked, the chief of the Mercury medical team, Air Force Lt. Col. Stanley White, said it would be "extremely hazardous" to abandon the Enos flight at such a critical proving time. "The MA-5 mission is more than a matter of just checking the spacecraft," he stated. "So far we have had experience with just one [unmanned] Mercury shot in orbit, and that for only one trip around the world. We need another shot now, for three orbits, so we can be sure that everybody in the system will have a chance to do his job."[6] In a worrying statistic, two of the four previous Atlas launches had ended in failure, so a great deal of work was still needed to make the Atlas a safer and more reliable launch vehicle.

On 29 November 1961 the MA-5 launch finally took place from Launch Complex 14. Enos had been woken at 3:00 a.m. and supplied with his normal breakfast. Then, dressed in his diapers and nylon flight suit, he was led out to a panel van which transported him around to Hangar S. Once there, he was given a thorough veterinary examination and clearance to fly by physician Capt. Dan Mosely. The handlers then escorted Enos out and into a waiting, air-conditioned medical van which would carry him out to the launch gantry, in much the same procedure as his astronaut counterparts. Once inside the van, sensors were taped onto his skin and a balloon catheter was carefully inserted. Enos was then placed into his custom couch and securely fastened down, although his hands were left free in order to allow him to perform his in-flight tasks. The lid was then closed on his container and sealed shut, the capsule's life support systems activated, and connections made to enable physiological data to begin flowing. The container was then carefully lifted out of the medical van, carried up in the gantry elevator and locked into place inside the Mercury spacecraft, ahead of a planned 7:30 a.m. lift-off.

The countdown was held at T-30 minutes, allowing technicians to reopen the hatch of the Mercury spacecraft and correctly position an on/off telemetry switch, causing an 85-minute delay. In the final two minutes before lift-off, Enos began systematically pulling his levers in response to the signal lights, which had now been activated. After several minor technical hitches had further delayed the launch, Enos was finally blasted into the skies at 10:08 a.m. During the ascent phase he was pressed back into his contoured couch with a maximum 7.2 g's. But he had often experienced this force pressing down on his body during centrifuge rides in training, and so knew it was only a temporary discomfort. Atlas 93D propelled the Mercury spacecraft up and out to the northeast. Once the booster rocket had done its task it dropped away and the capsule finally settled into an elliptical west-to-east orbit of 99 miles by 147 miles, inclined at an angle of 37 degrees to the equator.

All the spacecraft systems functioned as planned during the early part of the flight, and Enos happily resumed his tasks. In the first pre-programmed test period he won 13 banana pellets on the 50-count lever, and drank just under a liter of water through a tube. A small wiring malfunction on one of the tests meant that, like Ham, he received some undeserved electric tingles to the soles of his feet, but for the most part Enos did everything that was required. Despite an overheating problem, he remained calm. Everything seemed quite natural to him after his extensive training, and he continued jiggling levers and pushing buttons as his Mercury spacecraft swooped around the Earth. Occasionally he would relax

The container holding Enos in his flight couch is gently lifted from the transfer van at the launch pad ready to be inserted into the Mercury capsule. Handler Ed Dittmer is on the right of the photo. (Photo: NASA)

during pre-allocated rest periods, which had also featured in his training. His handlers were pleased to note that Enos did not attempt to hit any levers during these periods.

As the flight continued, a pre-recorded message was relayed from within the capsule to simulate voice contact with an astronaut. "CapCom [Capsule Communicator], this is astro," said the taped greeting. "Am on the window and the view is great. I can see all the colors and can make out coastlines." Later, at a post-flight press conference at the White House, an amused President Kennedy would tell reporters that "the chimpanzee took off at 10:08. He reported that everything is perfect and working well."[7]

A TROUBLESOME FLIGHT

The wiring defect that eventually gave Enos a recorded total of 76 highly irritating tingles to his feet proved to be just one of a series of problems detected when the capsule flew into the listening cone of Muchea in Western Australia during the second of its planned three orbits. Data emanating from the craft indicated there was a small but persistent wobble in the motion of the craft, while the attitude control system began to exhibit a degraded performance due to a thruster failure. The environmental control system was also unable to maintain the required temperature within the spacecraft. As a result, Enos's body temperature gradually rose to a dangerous 100.5 degrees Fahrenheit before it finally stabilized. The malfunctioning attitude control system eventually caused the spacecraft to begin slowly tumbling. This gave rise to considerable concern. There were grave fears that a lack of attitude control, combined with excessive thruster and propellant usage, could result in the capsule not achieving the correct position for retrofire. It was later discovered that the thruster problem resulted from a stray metal chip clogging a fuel supply line, which caused the spacecraft to drift from its planned attitude.

Flight Director Chris Kraft was worried that there might be insufficient fuel remaining at the end of a third orbit to achieve correct attitude control during re-entry. He consulted with his Mission Control team, and it was decided to end the flight early by setting procedures in motion to fire the retrorockets. The decision was made and implemented just twelve seconds before they would otherwise have had to commit to a third orbit.

Meanwhile, an overheated Enos was growing increasingly annoyed. In line with his training, he had been pressing the psychomotor buttons in the correct sequence, but the wiring defect meant he was only being rewarded by shocks to the soles of his feet. He began banging away at the buttons in frustration. Now in a state of outrage, he turned his attention to a different source of annoyance, grasping his internal balloon catheter and ripping it out. This action must have really hurt, but it seems he was beyond caring. Perhaps in reaction to the pain – or as a protest against all that had happened to him – he began fondling himself in front of the camera.

With the decision made in Mission Control to shorten the flight by an orbit, the necessary commands were relayed up to the spacecraft. The capsule slowed under the influence of retrofire, then dipped back into the atmosphere, heading for the planned recovery area some 200 miles south of Bermuda. The crew of a Martin P5M Marlin search aircraft eventually spotted the craft descending under its main parachute at around 5,000 feet. They relayed its position to the recovery destroyers USS *Stormes* (DD-780) and *Compton* (DD-705), which were 30 miles away. An hour and a quarter later the MA-5 capsule was safely plucked from the sea by the crew of the *Stormes*.

Once the capsule had been secured on deck, the hatch was explosively blown, the sealed container removed and opened, and an excited but overheated chimpanzee extracted from his couch. The temperature inside his airtight capsule measured 106 degrees Fahrenheit, but Enos soon cooled down and rapidly devoured two oranges and two apples. As a NASA report later recorded, "The subject had broken through the protective belly panel and had removed or damaged most of the physiological sensors. He had also forcibly removed the urinary catheter while the balloon was still inflated."[8]

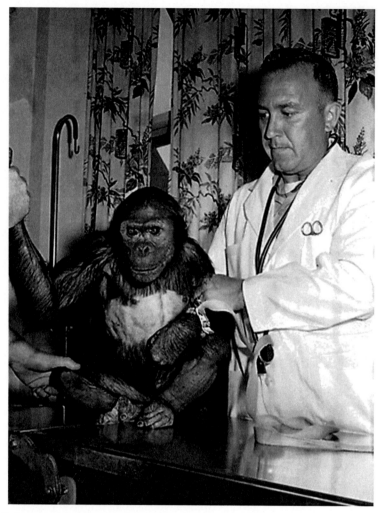

Post-flight, Enos undergoes a thorough veterinary examination by Capt. Jerry Fineg at Kindley Air Force Base, Bermuda. (Photo: NASA)

According to the official history of Project Mercury in the NASA publication *This New Ocean*, the USS *Stormes* dropped Enos at Kindley Air Force Base hospital, Bermuda, where chief veterinarian Capt. Jerry Fineg conducted a full evaluation of the animal's health and post-flight condition. As he reported, "the chimp was walked in the corridors and appeared to be in good shape. His body temperature was 97.6 degrees; his respiratory rate was 16; his pulse was 100. Apparently re-entry, reaching a peak of 7.8 g, had not hurt him."[9]

On the first day in December, Enos was transported by airplane back to the Cape. There he underwent a further battery of physical tests. The following week he made a joyful return to the chimp colony at Holloman Air Force Base in New Mexico.

Back at the Cape once again, Enos is escorted down the aircraft steps by M/Sgt. Ed Dittmer (right) and Airman Michael Berman. (Photo: NASA)

RESULTS PAVE THE WAY

The flight of Enos, and the way that he methodically performed his tests in 181 minutes of weightlessness despite the technical problems, gave an assurance that a human should be able to conduct any tasks which would be required of them in Earth orbit. Although troubling at the time, the issues affecting Enos's flight were able to be resolved, paving the way for John Glenn to make America's first manned orbital flight just three months later.

According to Dr. James Henry, the respective flights of Ham and Enos had demonstrated several crucial factors:

1. Pulse and respiration rates, during both the ballistic and orbital flights, had remained within normal limits throughout the weightless state. The effectiveness of the animals' heart action, as evaluated from the electrocardiograms and pressure records, was also unaffected by the flights.
2. Blood pressures, in both the systemic arterial tree and the low-pressure system, were not significantly changed from pre-flight values during three hours of the weightless state.
3. The two primates' performance of a series of tasks involving continuous and discrete avoidance, fixed ratio responses for food reward, delayed response for a fluid reward, and solution of a simple oddity problem, was unaffected by the weightless state.
4. Primates trained in the laboratory to perform during the simulated acceleration, noise and vibration of launch and entry were able to maintain performance throughout an actual flight.

Furthermore, Dr. Henry's project group was able to draw the following conclusions:

1. The numerous objectives of the Mercury animal test program were met. The MR-2 and MA-5 tests preceded the first ballistic and orbital manned flights, respectively, and provided valuable training in countdown procedures and range monitoring, as well as recovery techniques. The bioinstrumentation was effectively tested and the adequacy of the environmental control system was demonstrated.
2. A seven-minute (MR-2) and a three-hour (MA-5) exposure to the weightless state were experienced by the primates in the context of an experimental design which left visual and tactile references unimpaired. There was no significant change in the animals' physiological state or performance as measured during a series of tasks of graded motivation and difficulty.
3. The results met program objectives by answering questions concerning the physical and mental demands that the astronauts would encounter during space flight, and by showing that these demands would not be excessive.
4. An incidental gain from the program was the demonstration that the young chimpanzee can be trained to be a highly reliable subject for space flight studies.

"It was quite clear that the space effort at the beginning had to take the approach of a great expedition of exploration and adventure, and that research requirements should wait until the engineering problems had been solved," Dr. Henry told space researcher and author Shirley Thomas several years later. "In view of this, the Mercury animal flights were, in my opinion, an unexpectedly elegant and complex piece of combined physiological and psychological experimentation."[10]

A "GO" FOR ORBITAL FLIGHT

At a press conference held at the Cape press site after the recovery of the spacecraft and Enos by the USS *Stormes* south of Bermuda following the relatively successful two-orbit flight of the MA-5 mission, MSC officials expressed their satisfaction with the way it had gone. MSC Director Robert Gilruth remarked, "I would say we had a superb performance exhibited today on the part of all the various teams and on the part of the equipment. This includes the Atlas boost to orbit, the Atlantic Missile Range support, the [tracking] network and the network teams, the spacecraft, the checkout teams, the manufacturing teams, and the Navy recovery forces. I would also like to say that the fact that we decided to terminate the flight at the end of the second orbit lost very little of the spacecraft and other scientific data that we were after."[11]

Gilruth had already announced a few days earlier at NASA Headquarters in Washington that the next Project Mercury orbital flight would be a manned effort. He said that based on all available data, including a preliminary analysis of MA-5 information, it appeared that no further animal or unmanned flights were necessary. An analysis of the flight performance and post-flight physical condition of the chimpanzee Enos, together with a detailed study of the spacecraft, booster, and tracking network operations, had confirmed that the Mercury-Atlas system was now ready to proceed with the next orbital flight.

Speaking at the press conference a little later about the launch, Walter Williams, the Associate Director of MSC and Flight Operations Director for Project Mercury, said, "The boost was about as near perfect as you would expect to see."

Near the end of the conference Robert Gilruth announced that two crew teams had been selected for Project Mercury's first two manned orbital space flights. He named John Glenn as pilot for the first flight, with Scott Carpenter serving as his backup. Both men were there, participating in the conference. As well, Alan Shepard would act as pilot technical adviser for the MA-6 mission, and Gordon Cooper would head up the pad emergency escape and launch area recovery team. Gilruth then announced Deke Slayton as the

Flight Operations Director Walt Williams (second from right), astronauts John Glenn and Alan Shepard outside the Mercury Control Center with an unnamed program meteorologist. (Photo: NASA)

designated pilot for the second manned orbital flight with Wally Schirra as his backup. Gus Grissom would act as technical adviser for this flight.

Choosing his words carefully, Gilruth prefaced his announcement by pointing out that "this statement does not mean that we are necessarily not going to make other flights before a manned flight, nor that we are necessarily going with another flight this year."[12]

Although he had known of his assignment for some time, an obviously delighted John Glenn then indicated his pleasure over his selection. In answer to a press query he stated that he was set to go between then and the end of the year if the Atlas rocket was ready to go and the word was given.

Never given anywhere near the same acclaim as his illustrious predecessor Ham, Enos died less than a year later after his flight, having contracted dysentery from shigellosis, a bacterial infection in the lining of the intestines. Ham died in 1983 at the age of 26 and his soft tissue was buried with honor and a stone tablet outside the New Mexico Museum of Space History. Sadly, it is not known what became of the remains of pioneering space chimpanzee Enos.

REFERENCES

1. Carpenter, S., Cooper, Jr. L, Glenn, Jr., J., Grissom, V., Schirra, Jr., W., Shepard, Jr., A., and Slayton, D., *We Seven*, Simon and Schuster Inc., New York, NY, 1962
2. Telephone interview conducted by Colin Burgess with Edward C. Dittmer, Sr., 21 June 2005
3. *Ibid*
4. Alamogordo Space Center Oral History Program interview with Edward Dittmer, conducted by center curator George M. House, 29 April 1987, Alamogordo, New Mexico
5. Guenter Wendt interview with Peter Kerasotis for *Florida Today* newspaper, issue 29 October 1998, Pg. 12A
6. Clyde Bergwin and William Coleman, *Animal Astronauts: They Opened the Way to the Stars*, Prentice-Hall, Englewood Cliffs, NJ, 1963
7. Chris Kraft, *Flight: My Life in Mission Control*, Penguin Putnam Inc., New York, 2001
8. James P. Henry and Ed Mosely, *Results of the Project Mercury ballistic and orbital chimpanzee flights*, NASA, SP-39, Wash, DC, 1963
9. Loyd Swenson, James Grinwood and Charles Alexander, *This New Ocean: A History of Project Mercury*, NASA SP-4201, 1998
10. Shirley Thomas, *Men of Space* series (Vol. 7, chapter on James P. Henry), Chilton Company, Philadelphia, PA, 1965
11. NASA *Space Roundup News*, "Glenn, Slayton Are Named As Orbital Mission Pilots," Manned Spacecraft Center, Houston, Texas, issue Vol. 1, No. 4, 13 December 1961
12. *Ibid*

3

Marine on a mission

John Herschel Glenn, Jr. was born in a white clapboard house situated at 1525 Foster Avenue in Cambridge, Ohio. He came into the world on Monday, 18 July 1921. His parents, John Herschel and Clara (Sproat) Glenn, had both grown up in historic Cambridge, located in the foothills of the Appalachian Mountains in southeastern Ohio; a city internationally recognized for the manufacture of quality glass products.

THE GLENN FAMILY

John Glenn (Snr.) was mostly known to his friends and Cambridge locals as "Herschel." He could trace his ancestry back to Scotland and the MacGlynns, an offshoot of the MacKintosh clan. For 500 years the motto of the Glenn family has been *Alta Pete*, or "Aim High." A Glenn family researcher, Joe Boyes of Burbank, California, proved that some 200 years ago the Latin motto had been extended by two words to something remarkably prophetic for baby John Glenn, when it became *Alta Pete ad Astra* – which translates to "Aim High for the Stars."

Two years after Glenn's parents met, his father – who came from a farming family – enlisted in the U.S. Army. On 25 May 1918, prior to boarding a troopship bound for the battlefields of France, John and Clara were married. He subsequently fought on the Western Front as a member of the American Expeditionary Forces, with occasional duties as a bugler, but mainly delivering artillery shells to the front on trucks and horse-drawn caissons. During one battle a cannon shell exploded nearby; he was not wounded, but returned from the war with his hearing permanently impaired. He would require the use of hearing aids for the rest of his life. Returning to Cambridge, he first took on work as a fireman with the Baltimore and Ohio railroad, and then joined a local plumbing firm. In 1923, following the birth of John Jr., the Glenns moved from Cambridge to nearby New Concord, Ohio, a quiet, small Midwestern town where the senior Glenn set up his own plumbing and heating business at the corner of Main and Liberty; the Glenn Plumbing Company.

© Springer International Publishing Switzerland 2015
C. Burgess, *Friendship 7*, Springer Praxis Books, DOI 10.1007/978-3-319-15654-5_3

Clara (née Sproat) and John Glenn Snr. at Camp Sheridan, Alabama, 1918. (Photo: John Glenn Archives, The Ohio State University)

In fact the town was so quiet, a resident was once quoted as saying, "The top three topics of barber shop conversation are the weather, the weather, the weather."[1]

Glenn's mother Clara (from whom he inherited his red hair) had graduated from Cambridge High School and Muskingum College and became a schoolteacher, initially at local rural schools. She later taught at elementary schools in the Cambridge school system, and would meet her future husband while attending services at the East Union Presbyterian Church outside of nearby Clayville.

The Glenn family moved to this house in New Concord. In 1949, due to road works, it had been moved from its original site on the National Highway. (Photo: Author's collection)

John Glenn Jr., 9 months old. (Photo John Glenn Archives, The Ohio State University)

Clara Glenn gave up teaching soon after their baby boy was born and went to work instead at the plumbing store. As he grew up, young John spent a lot of time in the store, and as he mentioned in his later memoir, he even took naps swathed in blankets in a porcelain display bathtub.

The house the Glenn family had moved into was located on the National Highway (Route 40), perched on a bluff above the town's 'S' bridge. Many years later, in 1949, part of the bank in front of the house was cut away in order to widen the highway, and the house was moved up to the top of a hill on the Upper Bloomfield Road – later still renamed Friendship Drive in honor of the town's most famous citizen.[2]

John Glenn was still only a toddler when he first met the girl who would one day become his wife. Soon after the family moved to New Concord, his parents had struck up a lasting friendship with the nearby Castor family. Homer – a dentist by profession – and Margaret Castor were also active Presbyterians who had moved into the neighborhood around the same time from Columbus, Ohio. They had a cute baby daughter named Anna Margaret (always known as "Annie"), who was born a year before John. Together with three other young couples – Don and Glenna Cox, John and Hazel Sims, Charles and Helen Morehead – the Glenns and Castors formed a friendship group they called the "Twice Fives Club." Once a month they would all get together for a pot-luck supper at someone's house. Everyone brought their children along because, according to Annie Glenn, "nobody had baby sitters back then."[3]

As Clara Glenn once recalled, "When our children were small we generally took them with us when the club met. Annie and John weren't in school yet, and pretty soon they became playmates. When they started school, they just naturally 'paired off' whenever they went to children's parties."[4]

According to Glenn, "I don't remember when I first met my wife, Annie. Our parents were good friends and we practically grew up in the same playpen. We never knew a time when we didn't know each other."[5]

The Glenns were unable to have any more children. Instead, five years after the birth of John Jr., a baby sister named Jean was adopted into the family. By this time, to distinguish between the two Johns in the family, the younger Glenn had been given the nickname "Bud." He loved being around his father, partly because they shared a curiosity about mechanical things. "He always wanted to learn about new things, and he would go out of his way to investigate them," Glenn later wrote in his autobiography. "Although the Glenn Plumbing Company grew into a successful business of which he was very proud, I think he recognized the limitations of his education. He wanted to give me the curiosity and sense of unbounded possibility that could come from learning."[6]

FIRST FLIGHT

A strong and happy family, the Glenns managed to live and work through the tough times of the Great Depression. The plumbing business faltered (but ultimately survived), although Glenn's father had to take on some tough, additional work selling Chevrolet

John Jr., 4 years old, in a sailor suit. (Photo: John Glenn Archives, The Ohio State University)

automobiles to bring in extra money. Young John would often watch in boyhood fascination as his father tinkered with the motors of trade-in vehicles and would often help out, developing a deep and enduring interest in the way machines worked. Driven by an urge to help out his family financially, as well as saving up to buy a bicycle, he washed cars and sold rhubarb grown in produce gardens which the family cultivated behind the store and their home. Once he had saved enough money to purchase a used bicycle – it cost him 16 dollars – he also took on a job delivering the *Columbus Dispatch* newspaper.

As his mother once told a press reporter, young John always had a fascination for airplanes, and even before he began school he had been building makeshift "planes" by nailing bits of wood together and pretending to be a pilot. His father would often take him to the Port Columbus airport in order to spend some happy hours watching airplanes taking off and landing. Then came a truly momentous day in his young life, as he explained to some teenage reporters.

"When I was about eight years old, a man came to the little Cambridge, Ohio airport – it was a grass field – in a two-wing airplane with open cockpits. He was taking people up for rides. My dad and I were in our pick-up truck and driving by but stopped to watch. My dad asked if I wanted to go up. I'd always felt the ultimate would be to fly, so we went up, the two of us, strapped in the back seat of this open cockpit airplane. Once I had a chance to see things from an airplane, I was hooked. I'd built model airplanes, not the plastic ones you guys have now that click together, but the old balsa ones, where you had to carve every little piece and put it together with glue and cover it with tissue paper and shrink the paper, with a rubber band wind-up engine on the thing. I used to fly those and test different kinds, but I never thought I'd be able to fly, even though I thought that was something I'd love to do."[7]

Once he entered New Concord High School in 1938, Glenn joined in organized sports, eventually lettering in three sports: football, basketball, and tennis. In addition to becoming junior class president, he also played the trumpet in the band, was on the student council, was a reporter for the school newspaper, and performed in school plays. He would mostly excel at everything he tackled, but when asked said he was a fairly typical student back then, and nothing unusual.

"I studied hard and made pretty good grades. My mother had been an elementary school teacher, so there was always a lot of emphasis on education. My dad only had a sixth grade education, so he was self-taught, and did a lot of reading. He too thought education was important …. I had mostly A's with some B's and once in a while a C."[8]

Although he remained fascinated by airplanes and aviation, Glenn had not yet begun to think of it as a possible career. By this time the childhood friendship he had long shared with Annie Castor had grown and blossomed into love. They knew they would always be together. They seemed destined to marry one day and set up a home in New Concord, where Glenn would probably join his father's plumbing business. "Although Annie and I had put aside our plans for a teenage wedding in Kentucky, we both were certain that we would be married. It wasn't a question of if, only when. But at my high-school graduation in 1939, I was looking forward to college."[9]

The next academic venue for Glenn was Muskingum College, a local Presbyterian liberal arts school. His family had not fully recovered from the effects of the Depression and so he lived at home in order to save money.

"Muskingum is a Presbyterian liberal arts college," he would later write for the book, *We Seven*. "I told the professor who interviewed me before my freshman year began that I was interested primarily in technical work. My major was chemical engineering. I was still not interested in aviation as a profession."[10]

John and Annie both attended Muskingum College – even though Annie could have taken up a scholarship she won to Juilliard School in New York – and while she majored in music, John studied chemical engineering.

John Glenn and Annie Castor. (Photo: John Glenn Archives, The Ohio State University)

Then, in January of 1941, he read that the U.S. Department of Commerce would pay for a Civilian Pilot Training Program, which included flying lessons and college credit in physics.

"Well, in my junior year in college I saw a notice for the Civilian Pilot Training Program – the government was sponsoring it. If you signed up, you could get flight training, your private pilot's license and credit for physics because we were studying engines and thermodynamics. I thought that was too good to miss; my dad wasn't much for this, he saw aviation as a little too dangerous. But gradually he came around, and that's when I started flying. I loved it, and I still do to this day."[11]

The pilot training program took place on a grass airstrip at the New Philadelphia airport, and he would drive out in a borrowed car along with fellow students Jack Cox, Dick Atchinson and Dane Handschy. They had soon learned the basics of flying, both in the classroom and in the cockpit of a Taylorcraft BL-65 trainer, a side-by-side seating, high-wing monoplane. His instructor, Harry Clever, once wrote in his flight recorder that Glenn was "eager to learn, relaxed, alert and [showed] good coordination."[12]

He would receive his pilots' license and save enough money to hire small airplanes to maintain his proficiency and love of flying.

JOINING THE MARINE CORPS

The savage and unprovoked attack on Pearl Harbor would have an immediate impact on any future plans for John Glenn and Annie Castor.

Glenn discussed the day he was driving to Annie's senior recital, where she was to play *Be Still My Soul* on the organ at Muskingum's Brown Chapel. "She was and is an accomplished musician who was offered a scholarship to study at Juilliard in New York. It was Sunday, December 7, 1941. A bulletin came over the radio. The Japanese had bombed Pearl Harbor."

Shocked by the appalling news, Glenn sat through Annie's recital, hardly hearing it as he pondered the implications of what he had heard on the radio, and how it might affect his life and his future with Annie. Once the recital was over they talked about what he should do and reached an understanding. Just a few days later he enlisted in military flight training, with a wedding to follow.

He completed pre-flight training at the University of Iowa and primary training at the Naval Reserve Base in Olathe, Kansas, completing his flight training in Corpus Christi, Texas. He was commissioned a second lieutenant in the U.S. Marines Corps and also graduated as a naval navigator on 31 March 1943. A day later, having been awarded his wings, Glenn took a train home to New Concord, with a 15-day leave pass in his pocket. It would take him three days to get there. He and Annie had set a wedding date based on his return home after the graduation ceremony, and all the wedding preparations had been taken care of in his absence. "I didn't have to do anything but show up," was his bemused recollection.

On 6 April he and his beloved Annie were married at the College Drive United Presbyterian Church. As he had to report again for duty three days later they had a brief but romantic honeymoon, spending their first days as husband and wife in Columbus, Ohio.

To the delight of the newlyweds, Glenn's father gave them a black 1934 Chevrolet coupe from his used car lot as a wedding present. They packed everything into the car and after a series of mechanical problems and flat tires they finally made it to the Marine air

John Glenn as a U.S. Navy cadet, 1942. (Photo: John Glenn Archives, The Ohio State University)

station at Cherry Point, North Carolina, where a whole new life would begin for them, and a whole new career would open up for John Glenn.

In October that year he was promoted to full lieutenant. After some advanced training in Camp Kearny Mesa, near La Jolla, California, he was assigned to VMJ-353, a transport squadron flying R4Ds – a Navy and Marine version of the renowned workhorse, the Douglas DC-3. This was followed by a move south to the Marine Corps Air Station at El

Newlyweds John and Annie Glenn. (Photo: John Glenn Archives, The Ohio State University)

Centro, near the Mexican border. Here he had his first experience with a fighter, undertaking gunnery practice in F4F Wildcats with VMO-155 Marine squadron.

On 6 January 1944 Glenn received his embarkation orders, and the following month the squadron was shipped out to Hawaii to await further orders. On 28 June they sailed for the Marshall Islands, where Glenn spent the next year flying 57 strafing and bombing missions in a Chance Vought F4U Corsair fighter-bomber with VMO-155 in the vicinity of the islands. Depending on the targets assigned to the squadron, the Corsairs were often loaded with

500-pound armor-piercing bombs designed to penetrate concrete bunkers used by the Japanese to store fuel and ammunition. "You knew you were on target if you got a secondary explosion after the first hit," Glenn once reflected. "The secondaries could be spectacular, as well as good news to our intelligence officers in the debriefings at the end of every flight.

"We would go through spells of several days in which hardly any antiaircraft fire came at us. Then one day we would fly over the same island and they would throw everything at us but the kitchen sink. We would fly through clouds of tracers so thick it was hard to see how a plane could make it through without getting shot full of holes, but usually we did."[13]

Lt. Glenn returned from the Marshall Islands campaign with two Distinguished Flying Crosses and ten Air Medals, and in July 1945 was promoted to captain. At war's end he

Standing in front of an F4U Corsair, Marshall Islands. (Photo: John Glenn Archives, The Ohio State University)

Annie and John Glenn with his parents, Clara and John Snr., in 1945. (Photo: John Glenn Archives, The Ohio State University)

was still unsure whether he would be offered a regular commission in the U.S. Marines, which didn't really trouble him at the time. Even though he liked the Marines and their *esprit de corps*, he still didn't know if he wanted to make the service his career. "What I did know was that I loved flying and wanted to keep doing it somehow."[14] His father was disappointed, however – he had wanted his son to join him in the plumbing business.

Glenn was then stationed for a time back at Cherry Point, North Carolina, and later did a tour of duty conducting test work at Patuxent River, Maryland, where he was engaged in putting brand-new airplanes through simulated combat missions as part of the Accelerated Service Test. Afterwards he was assigned to El Toro, California, helping to train other pilots. It was during this time that his commission finally came through. He discussed it with Annie, but she knew by now how much flying meant to him and said she was fully supportive if he decided to stay in the Marines. He accepted the commission ... and never looked back.

On 13 December 1945 John and Annie became the proud parents of a son they named John David Glenn.

PATROL DUTY IN CHINA

After Cherry Point, Glenn was assigned in March 1946 to El Toro in California as the operations manager for Marine Fighter Squadron VMF-323, known as the "Death Rattlers." He found life at El Toro was good, but somewhat unsatisfying. "I didn't seem to be learning anything new as a pilot," he once wrote. "It felt as if I were treading water while the postwar military was trying to figure out what to do with itself."[15] Then he heard that the Marine Corps was looking for volunteers to take on what was described as "a short tour of duty" in China. He decided to volunteer for the experience that it would give to his flying.

Glenn arrived at Nan Yuen Field, south of the Chinese capital (then called Peiping), in December 1946 to join the Corsair squadron VMF-218 as acting operations officer. He spent the next two years overseas flying Corsairs on patrol duty in North China and later Guam with the squadron. As the squadron was preparing to move on to Guam he got the welcome news from home that Annie had given birth to their second child, a girl they named Carolyn Ann, on 19 March 1947.

Glenn would return home in December 1948, two years after his so-called "three-month tour" overseas had begun, and was finally reunited with his now-expanded family. Once he had settled in again, he spent almost three years at the Naval Air Training Center at Corpus Christi, Texas, firstly as a student at the Naval School of All-Weather Flight, and then as an instructor at the school.

His tour in Training Command at Corpus Christi came to an end in the late spring of 1951. He and Annie put their house on the market and made plans to move back east to Virginia, where he had been ordered to the Marine Combat Development Command at Quantico, in order to attend a course in amphibious warfare. He then stayed on at Quantico, serving briefly as the assistant operations officer for its air station, Turner Field.

In July 1952, Glenn was further promoted to the rank of major, and then went through a jet fighter refresher course at Cherry Point before returning to active service in the Korean War, flying a Grumman Panther jet with VMF-311.

WAR IN THE AIR OVER KOREA

During his Korean service, in which he flew 90 missions between February and September 1953, Glenn was awarded his third and fourth Distinguished Flying Crosses and a further eight Air Medals. His ground crew became used to Glenn bringing back his Panther jet full of holes. As one of them recalled, "He had a habit of getting right down on the deck on his runs. He always hit the target that way, but he sure caught a lot of flak doing it!"[16] On one occasion, his incredulous crew chief counted a total of 203 holes in Glenn's Panther. From then on, that airplane was christened "Glenn's flying doily."

In June 1953 Glenn was assigned to the 25th Fighter-Interceptor Squadron based at Suwon, which meant he would finally see some air-to-air combat. The squadron, equipped with F-86 Sabre jets, regularly patrolled along the Yalu River, which was known to the airmen as "MiG Alley." On 12 July, six days shy of his 32nd birthday, Glenn shot down his first enemy plane – a MiG-15 – in aerial combat. Just a week later he bagged his second MiG-15,

John Glenn at the U.S. Navy Test Station, Patuxent River, Maryland in 1954. (Photo: John Glenn Archives, The Ohio State University)

and was firing at another when he ran out of ammunition. His third and final kill came on 22 July. In fact the MiGs that his squadron scored on that day were the last three to be shot down during the Korean War, because a truce was declared just five days later.

When his Korean tour was completed, Glenn applied for duty at the Navy Test Pilot School at Patuxent River, Maryland (known simply to all U.S. military aviators as Pax River) and was accepted, entering the program in January 1954. After completing the school, he stayed on at Pax River, where he helped test many of the Navy's newest jets, fighters in particular, and became an F8U Crusader project officer at the Armament Test Division. He was later assigned to duty in Washington, with the Fighter Design Branch of the Navy's Bureau of Aeronautics, which he says gave him a chance to gain experience in the design of new planes and equipment and to get a pilot's viewpoint of many new design items.

Glenn in the cockpit of his Project Bullet F8U-1P Crusader. (Photo: John Glenn Archives, The Ohio State University)

On 16 July 1957, Maj. Glenn won a fifth Distinguished Flying Cross when he made the first non-stop supersonic trans-continental flight in a Vought F8U-1 Crusader in what was called Project Bullet, setting a new record in flying from Los Angeles to Floyd Bennett Field, New York, in 3 hours 23 minutes and 8.4 seconds, at an average speed of 723 miles an hour, or 63 miles an hour faster than the speed of sound at 35,000 feet. He had beaten the existing record by a margin of 21 minutes.

"I had crossed the country in considerably less time than the project's namesake bullet would have taken, and the record now belonged to the Navy and Marines," he would later proudly recall.[17]

After the record-breaking transcontinental flight, a happy Glenn poses with his family. (Photo: John Glenn Archives, The Ohio State University)

PREPARATION FOR SPACE

In October 1957, two notable events occurred in the life of John Glenn. Three months earlier he had broken the trans-America speed record flying a Crusader aircraft, and this touch of celebrity led to an invitation to make his first television appearance on a popular Tuesday evening CBS game show called *Name That Tune*.

Maj. Glenn, in his dress uniform with a chestful of medals, was paired with a personable young boy in a Cub Scout uniform named Eddie Hodges. Over five weeks they took part in the music quiz show, and shared in $25,000 as the winning team.[1] The second major event was the launch into orbit of the first Soviet satellite, *Sputnik*, on 4 October 1957.

In November 1956, nine months before his record flight in Project Bullet, and after nearly two years at Patuxent River, Glenn had been transferred to Washington, D.C. as a project officer in the Fighter Design Branch of the Navy Department's Bureau of Aeronautics. He was still there in early 1958 when his office was asked to furnish a test pilot with extensive flight experience to visit the laboratory of the National Advisory Committee for Aeronautics (NACA) at Langley Air Force Base in Virginia to make a few runs on one of the space flight simulators as part of its investigation into different shapes and characteristics of a number of re-entry shapes. The same officer would then proceed to the Naval Air Development Center at Johnsville, Pennsylvania, undergoing runs on their large centrifuge in order to compare data obtained from the Virginia simulator with that obtained from the Johnsville centrifuge while under high g forces. Glenn volunteered for the assignment and was accepted.

---

[1]   For a full account of their participation in *Name That Tune* and how it featured in the film *The Right Stuff,* see Appendix 4.

The Glenn family in 1957. (Photo: John Glenn Archives, The Ohio State University)

"I spent a few days at Langley and more than a week at Johnsville, where I helped work out a mission on the centrifuge that simulated the conditions a pilot would get into as he made a re-entry from space. It was an interesting taste of the future."[18]

Soon after, he was handed another space-related assignment at the McDonnell plant in St. Louis, Missouri, where they were working on the development of the blunt-ended Mercury space capsule that might one day carry a pilot into space. Meanwhile, preparations had begun that would one day give John Glenn a far closer association with that capsule than he could ever have envisaged.

NASA IS BORN

On 2 April 1958, in response to Soviet space efforts that were proving demoralizing to the American public, President Eisenhower sent to Congress a bill calling for the immediate establishment of a civilian aeronautics and space agency. Congress duly passed the Space Act on 29 July, and NASA officially came into existence on 1 October.

A month later, on 5 November, a Space Task Group was formed at the Air Force's Langley Research Center with Robert Gilruth appointed as its director. On behalf of NASA, this task group was given four major objectives: to prepare specifications for a manned spacecraft; to plan and build a worldwide tracking network; to select and develop a suitable launch vehicle; and to select and train potential space pilots who would undergo a two-year training program. In pursuing the latter objective, Charles Donlan, a senior management engineer and Assistant Director for Project Mercury, served as chairman of the astronaut selection committee. His committee comprised of aerospace engineers, flight surgeons, industrial psychologists, and psychiatrists from all branches of the military. They were to undertake the initial screening of records, and then carry out interviews and testing of the selected candidates on behalf of NASA.

With no precedents or government procedures to follow, NASA had to decide where the best candidates could be found, how many were needed, and how they should be tested. What they *did* know was that the selection process would hinge on three crucial factors: physical, psychological, and technical.

In the last week of 1958, after several meetings between NASA Administrator Keith Glennan, his deputy Hugh Dryden, the Space Task Group's Robert Gilruth, and other upper-level representatives from NASA and the Space Task Group, a consensus was reached. For speed and facility in arriving at the selections, it was decided to restrict the search to the ranks of military test pilots. There were several reasons for this: test pilots knew the rigors of military life, they were available at short notice, and their full service and medical records were available. First, however, the decision had to go before President Eisenhower. As Keith Glennan recalled, the President said, "Of course you should use service test pilots. They are in the service to do as the service requires of them at various times. They ought to have a chance to volunteer if they wish." As Glennan said, "We got it cleared in five minutes."

NASA Administrator Keith Glennan shows President Eisenhower images from the TIROS-1 meteorological satellite. (Photo: Dwight D. Eisenhower Library)

It was decided to carry out the medical testing at an independent medical facility in New Mexico called the Lovelace Clinic, while further stress testing and psychological evaluation would take place at the Wright Air Development Center in Ohio, which had already been involved in evaluation testing of space candidates for other potential service programs.

The criteria needed in a candidate were soon pounded out by the Space Task Group. A candidate needed to possess a university degree; be a graduate of a test pilot school; have around 1,500 jet hours; be in superb condition, mentally and physically; be no taller than 5 feet 11 inches, this restriction dictated by the confines of the Mercury spacecraft; and be less than 40 years of age. NASA had already decided that it would need twelve Mercury astronauts to fly the first missions.

By this time, despite official secrecy surrounding any manned space program, Glenn had begun picking up leaked information through his superiors at the Bureau of Aeronautics, as well as hearing strong rumors of a process to select qualified test pilots to train for the first pioneering space flights. He was keen to become a part of this program. However, he also realized that he was lacking in certain crucial qualifications.

"Thanks to the timing of the war, I was not a college graduate – although service schools and additional university work I had completed while I was in the service had given me the equivalent of a college degree. Also, at thirty-seven, I was probably a little older than most of the men NASA was considering. I was philosophical about my chances, however. I was a senior major in the Marine Corps, and if I were promoted to lieutenant colonel I would probably be in line to command a Marine air squadron. My career, then, was in pretty good shape. I did know that space travel was at the frontier of my profession, and I naturally wanted to be in on it. But the gratification and simple pride over the fact that I might not have enough brains and stamina and experience to be chosen was the least important of my reasons for wanting to try. I felt that many of my experiences had added up to prepare me for the kind of challenge that Project Mercury presented and that I would be remiss if I did not volunteer to put some of this background to good use."[19]

Like many other military test pilots, he would receive a summons early in 1958 to attend a top-secret briefing in the Pentagon, where everything he had heard about Project Mercury and the need for pilots – to be known as astronauts – was confirmed. However, he had come very close to missing out.

SELECTING THE MERCURY ASTRONAUTS

Earlier, behind the scenes, those members of the Space Task Group involved in the initial selection phase, or Phase One of the operation, had traveled to the Pentagon. There, using broad selection criteria, they had pulled the records of 508 candidates. The next step was to check their medical records and reports from their superior officers, ensure they had the minimum amount of jet flying hours, and also examine the type of flying they had been doing.

Out of 225 Air Force records screened, only 58 met the minimum standards. 225 Navy records were screened and only 47 made the grade. Of 23 Marine Corps records screened only 5 men met the minimum standards, and though 35 Army records were screened no one met the requirement of being a graduate from a test pilot school. Back then there were no female military test pilots, so women were excluded from consideration. Hence out of the 508 records that were screened, 110 officers met the minimum standards.

However the initial selection process was not without its problems, and many suitable candidates probably missed out through clerical errors. In fact one particular candidate was almost overlooked, as U.S. Navy psychologist Robert Voas later revealed.

"I was the Navy representative on the selection group and I made the error of assuming that the records of the Marine Corps pilots would be in the same files as the Navy pilots, since they went to the same Patuxent River training center. It wasn't until we had selected almost all the 110 candidates that it was called to my attention that we didn't have any Marines. So we had to make a quick contact with the Marine Corps and they found two pilots who fit the requirements. One was John Glenn."[20]

As Glenn had deduced, the selection panel would find one major problem with him, in that he lacked a college degree. "Frankly," Glenn told biographer Frank Van Riper, "I had more than enough credits for a college degree. In fact, I probably had enough for a masters, based on all the academic work at Patuxent, as well as college-level work I had done right after World War II through the Armed Forces Institute. Around that time, I had also gone to the University of Maryland extension division two or three nights a week at the Pentagon.

"I had transferred all this back to Muskingum [College], but they still wouldn't give me a degree. They held it up on a residency requirement, of all things! I'd only spent the first twenty years of my life there, and they stymied it on that."[21]

U.S. Navy psychologist Dr. Robert Voas with John Glenn. (Photo: NASA)

Dr. Voas recalls that this lack of a degree was downplayed by the selection committee in the case of Glenn. "From the beginning, it was felt that while John didn't have a formal degree, he had the outside course work, and that was the equivalent. We didn't take these requirements that rigidly. What was important was the nature of the flight record, and John's was outstanding."[22]

Glenn commented further on this in his 1999 autobiography, *John Glenn: A Memoir*, stating that his remaining in the candidate field was entirely due to Col. Jake Dill, his one-time commanding officer at the Marine detachment at Pax River and, at that time, second-in-command of personnel at Marine Headquarters in Washington, D.C. "As the field narrowed, my lack of a degree loomed larger. Jake Dill, unbeknownst to me, learned that I had been deselected on the basis of the degree requirement, since the astronauts were all to be assigned specific, and highly technical, tasks within the program in addition to riding rockets into space. Jake knew my background and thought this was unfair. He went to NASA with all my combat and academic records, and my technical flight reports from Patuxent. I later learned he had met with the selection board and convinced them that I had more than the equivalent of a degree."[23]

Once the candidate list had been narrowed down to 110, each was then ranked in terms of his overall qualifications and the reviews were then placed in ranking order, from the most promising down to least promising. For Phase Two of the operation, these men were to be called to Washington in three groups, starting with those most highly ranked, under secret orders and in civilian clothing to attend a briefing by a senior officer from their service, as well as NASA officials.

The first group of 35 candidates, including Maj. Glenn, turned up at the Pentagon on Monday, 2 February 1959. Here the Air Force candidates were initially briefed on Project Mercury and what it might mean for them service-wise by the Chief of Staff of the Air Force, General Thomas White, while the Navy and Marine candidates were simultaneously briefed in another room by the Chief of Naval Operations, Admiral Arleigh Burke. Up to this time most of the men knew very little about Project Mercury or what it entailed. In fact Glenn, through his previous work for the NACA and with McDonnell in St. Louis, was one of the better informed and better prepared candidates.

After their service briefings, the group were collected together in one room to receive a more specific NASA briefing and an outline of Project Mercury by Charles Donlan, Warren North, a NASA test pilot and engineer, and Lt. Robert Voas, the Navy psychologist. They were then told that if they wished to opt out of consideration at this time, this would not be held against them or noted in their service record.

Once they had indicated their willingness to continue through to the next phase, the candidates were subjected to an initial suitability interview by service psychologists, Drs. George Ruff and Edwin Levy, after which they sat through a review of their medical history. Some applicants would be found to be higher than the maximum 5 feet 11 inches and were eliminated from the process. Following the second round of briefings a week later, a total of 69 men had been processed, and with a higher than expected volunteer rate, Charles Donlan decided to cancel the third round of briefings, as he had more than enough suitable applicants to fill the twelve positions.

HOW THE SELECTION PROCESS WORKED

Responding to a query from the author in 2010, Lt. Col. Dr. Stanley C. White, who was part of the selection team as Chief, Life Sciences Division, agreed to offer his reflections and opinions. First and foremost, he wanted to emphasize that the selection process was not just a medical event.

"The medical effort was combined with the flight operation segment of the Space Task Group to arrive at the pool of candidates. The goal of the medical segment of the selection process was to offer a pool of acceptable candidates, health-wise, for the NASA selection team to pick from."

Initially, it was understood by the candidates and those administering the initial briefings in Washington that twelve candidates would be chosen, but as Dr. White explained, this number was decreased prior to the final selection.

"During the discussions of the selection team that followed, several decisions that would change the final results had to be made. It was decided to reduce the number to be selected from twelve to seven or eight. The issue leading to this number was the number of flights that were being planned versus the number who could be expected to get a flight."

Many of the candidates declined selection because they felt they would have little, if no control over their craft, but in reality this was not the case, according to Dr. White.

Dr. Stanley C. White. (Photo: NASA)

"Preparing the spacecraft design proposal, we put information and associated controls for the crew members to use in the flight operation rather than having him a passive rider in the flight. Our first flight goal was to prove this was a valid approach. There were many critics from all disciplines who felt this approach was foolish and would doom future space exploration. Plus they felt the weight this decision took [in adding control apparatus] could be better spent for other purposes. With this approach, the operations side of the Space Task Group began to identify the background and experience needed to answer the questions as to what we should test to prove one way or another the scope, capacity and quality of the astronaut's performance. That is how we got the thirty-one candidates to be in the final pool. It was from this pool that the final seven were chosen by the NASA selection team. It was a rich pool, since some of this final pool came back and were selected in later selection cycles.

"One other point; the Medical, Psychological and Physiological team worked on the concept that there was an abundance of candidates available, therefore we only offered up candidates that were judged fully acceptable from our point of view. If the operations side of the house had some reservation on one of the candidates, we re-examined his file and discussed it in detail with the operations staff."

When asked if the selection process held to a rigid format prepared ahead of time and then followed a timetable holding to this schedule in its execution, Dr. White responded:

"Quite the contrary – each part of the selection process was dynamically re-examined as we went along and revised as other events in the total program happened that could impact the selection process."

When asked for an instance of this, Dr. White offered a surprising piece of information.

"For example, as the design of the interior of the spacecraft became clear, we found that the important dimension for height of the crewmember was not the original specs on the total height of the standing candidate, but the butt-to-top-of-head height. As a result we had to revise our specs on what was judged a fully acceptable candidate. This later was reinforced when we tried to position the crewmember in the seat to allow him to use the periscope with his pressure suit helmet on and with his visor both closed and open. Thus, as the steps in the selection process were completed, some candidates became more attractive and others less so from a physical, physiological, and human factors point of view.

"At the same time, the operations side of the selection team was refining what tests they would conduct to demonstrate man's capability to first survive and then participate in the flight-related events as a way to quantify his performance capability. Example: we built an eye chart to proper scale to be viewed through the periscope as he passed over Texas [and] changed the positions of the symbols orbit-to-orbit to insure validity of what the crewmember reported. It proved conclusively that a good man could see as well as a trained recce pilot in an aircraft, even though he was many miles higher than the altitude of the recce pilot."

Asked about the candidates' individual responses to these particular tests, Dr. White replied:

"Different candidates responded to this environment of dynamic change differently; some liked the challenge, some did not; thus the changing of minds as they became more clear that they were a big part of the mission test program. Also remember these were all test pilots who came from a background of winged vehicles with engines which they controlled, and [launching] on a missile fired by somebody on the ground was considered a major change in their work environment. This issue caused several to not volunteer originally and some later as they thought about it. Some wanted to have a rigid set of detailed specs on the mission just as they had had in the military programs they had worked on."

Dr. White then re-emphasized that this project was dynamic and changing by the day.

"This bothered some enough to drop out later as they were exposed to this shift in events as they went through the selection series of events.

"The flight operations staff of the project were equally busy and were identifying what technical skills they were favoring for the crewmembers to bring to the flight operation. This made those with these skills more attractive than those without such skills. Such skills – recce experience for example – were considered in the final stages.

"All of these actions were directed toward offering to the final selection committee a fully acceptable pool of people from which they could choose. All who were in this final pool probably could have done an excellent job and were so advised. Those with minor and/or correctable problems were also advised of their problems and told to try again in the future.

"Keep in mind that this was one of the first projects which included the crew member as part of the issues to be studied from the start. To this time man had been considered a 'given' and not part of the study. In our case, man and his ability to survive was Question No. 1, and assuming success, followed closely by the study of his capabilities and ability to contribute to the success of the mission as Question No. 2. Most of these candidates had never been confronted with this environment."[24]

LOVELACE CLINIC AND WRIGHT-PATTERSON

Eventually, out of the 69 candidates who attended the briefings, six were judged to be too tall, and a further 16 declined to continue, leaving 47. Further checks and testing by NASA eliminated a further 15 candidates, bringing the number down to 32.

These men were then sent to the Lovelace Foundation for Medical Education and Research (better known as the Lovelace Clinic) in New Mexico, to undergo incredibly rigorous and often bewildering medical, physical and psychological evaluation. Dr. W. Randolph Lovelace II oversaw the medical and biological aspects of the selection process, designed to determine the physical health of each of the candidates and reveal any problems that could preclude them from traveling into space.

The first group of astronaut candidates entered the Lovelace Clinic on 7 February 1959, ready to face seven and one-half days and three evenings of intense, painful, humiliating and even degrading medical scrutiny and thirty different laboratory tests. These were divided into six basic components: the candidate's medical and aviation history; a physical examination; laboratory tests; a radiographic examination; physical competence and ventilator efficiency tests; and then a final evaluation. The competition would be fierce, and the examiners were under strict instructions to eliminate anyone who proved to be in less than perfect physical condition and health.

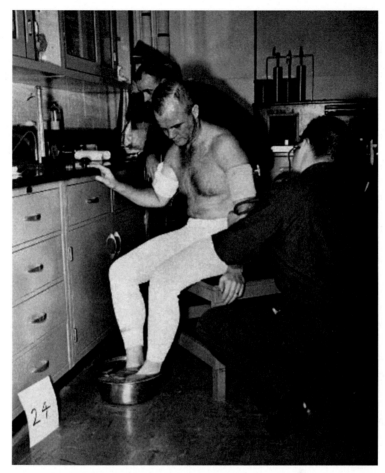

John Glenn undergoing stress testing at Wright-Patterson. (Photo: U.S. Air Force)

As Dr. Lovelace later revealed in an article for *Life* magazine: "We could not afford to overlook any test that might catch even a minor defect. There might, for example, be a tiny congenital opening between the right and left sides of the heart. Normally a man so afflicted might never show a sign of heart trouble, but under extreme circumstances, such as sudden decompression at high altitude, such a defect could mean death."[25]

James Lovell said in his autobiography *Lost Moon* that it was a "nightmare week," stating that, "In submitting to these whole-body violations, the candidate astronauts would have their livers injected with dye, their inner ears filled with cold water, their muscles punctured by electrified needles, their intestines filled with radioactive barium, their prostate glands squeezed, their sinuses probed, their stomachs pumped, their blood drawn, their scalps and chests plastered with electrodes, and their bowels evacuated by diagnostic enemas at the rate of up to six per day."[26]

"The medical tests were required not just to see who could withstand the rigors of it all," Dr. Voas later stated, "but also to ensure that, after we had put something like a million dollars into each of these guys, that we had a good long-term investment."[27]

Having recorded all their findings, the Lovelace physicians would finally release each candidate group for travel to the Wright Air Development Center (WADC), an auxiliary of the Air Research and Development Command (ARDC) of the U.S. Air Force at Wright-Patterson Air Force Base. In 1952 the Human Factors Division of the ARDC had begun a program specially aimed at selecting pilots for special research flights. Within the WADC was the Aero Medical Laboratory (AML) where, beginning on 16 February, the astronaut candidates would undergo psychological and stress tests.

This phase of the selection process involved a five-day battery of psychological, anthropological, physical fitness, biological acoustical, thermal and acceleration tests conducted by the AML investigating panel. The purpose of the Lovelace Clinic examinations had been to establish the medical status of the finalists, whereas the WADC program was to determine the physical and psychological well-being of each candidate. It was a critical series of examinations, due to the anticipated physical and mental stresses associated with space travel. John Glenn would be in the final group of six candidates, Group VI, which included a Navy pilot with whom he had forged a great friendship – Scott Carpenter.

After the examiners had conducted an extensive evaluation of each finalist's physical status the men were re-interviewed, during which psychological measures of their motivation and personality were re-examined. For the weary candidates it meant that after strenuous days filled with mind-sapping stress testing they would then be subjected to several hours of psychological grilling, severely testing their endurance, patience, and concentration. All of the tests were completed with the departure of the last group of finalists nearly six weeks later, on 28 March, and once the results had been assembled they were sent off to NASA, who would make the final cut, bringing the final number down to just seven.

In an oral history interview for NASA, Robert Voas admitted that Glenn's "cooperative attitude" at Lovelace (and later at Wright) had probably earned him some disproportionately high test grades from the clinic's examiners.

"The interesting thing about John is that he overwhelmed everybody. He had a charismatic personality, and just about everybody who dealt with him was highly impressed and rated John right up at the top. I always smile a bit about that because in some of the physical and mental tests, for example, his scores weren't all that much better than those of some of the other candidates. But the physicians gave him top evaluations due largely, I believe, to his strength of personality and dedication."[28]

On 1 April 1959, John Glenn was promoted to the rank of lieutenant colonel. Eight days later, at an announcement ceremony held in NASA Headquarters in Washington, D.C., he was named as one of seven test pilots who would henceforth be known America's Mercury astronauts.[2]

---

[2]   For a full description of the Mercury selection process, see the author's 2011 Springer-Praxis book, *Selecting the Mercury Seven: The Search for America's First Astronauts.*

The newly announced Mercury Seven astronauts. From left: Wally Schirra, Alan Shepard, Gus Grissom, Deke Slayton, John Glenn, Scott Carpenter, and Gordon Cooper. (Photo: NASA)

The evening of his announcement as a Mercury astronaut, John Glenn was photographed at his Arlington home with wife Annie, son David, 13, and daughter Lyn, 12. (Photo: Associated Press)

INITIAL TRAINING

Following an orientation period, the seven Mercury pilots took on flight-relevant tasks while working out of a barracks building at the Langley Research Center in Virginia. As planned, each was assigned to a different area of specialty within Project Mercury and during regular group discussions (which they came to call their "séance sessions") reported on their work and findings. In this way they could all benefit from work of the others without the need for each of them to be fully familiar with all of the hardware, systems, and principles and could concentrate within their own specific field of work.

As Deke Slayton recalled, "It was obvious to us from the start that Project Mercury was much too complex and far-reaching for all seven of us to learn everything there was to know about in every detail. If we had tried to do that – read all the stacks of technical manuals and go out and visit all of the hundreds of contractors and subcontractors who were building bits and pieces of the system – we'd still be in the classroom. So we split the work up between us .... Everything was parceled out fairly evenly. We each went our own way, dug into our subject, and then made reports back to the others wherever we discovered something we thought they ought to know."[29]

Scott Carpenter was involved in communication and navigational equipment; Gordon Cooper took on the Redstone rocket system and everyday procedures; Gus Grissom worked with McDonnell engineers on the hand control and autopilot systems of the spacecraft; Wally Schirra's responsibilities were the Mercury pressure suit and the environmental systems; Alan Shepard worked on Project Mercury recovery operations; and Deke Slayton became involved with the Atlas rocket and flight procedures. As John Glenn had previously worked in aircraft design, his specialized area was the layout of the instrument panel which they would also sit in front of within the Mercury spacecraft.

Wally Schirra described Glenn's work on the instrument and cockpit layout as "a real 'opinion area,'" by saying, "You can get as many different opinions about the layout of any cockpit as you have pilots who are going to sit in it. And you can probably never satisfy everybody completely. Basically, though, we all felt that some improvements could be made in the original layout .... John rode herd on all this to make sure that all seven of us knew exactly what was being done and that we were all more or less in harmony when the workmen were ready to start building the thing."[30]

"We also had to learn about our new machine," Deke Slayton added. "That was really the most important thing. All of us knew quite a lot about the vagaries of untested airplanes. But we knew practically nothing about missiles and rockets when we first arrived at Langley. We took it for granted that boosters could explode on the launching pad and that failures up in the air were everyday risks in the missile business. We also knew, after a lecture or two from Convair, that the big Atlas had about 40,000 parts in it, any one of which could conceivably go haywire at any second, even after you thought you were on your way. The spacecraft had several thousand more parts in it that could go wrong. But we also knew that any part in your automobile could act up – like the brakes when you are in the middle of an icy hill. We figured that the Atlas machine would work smoothly whenever we climbed in and stepped on the starter – provided the people who built it and aimed it knew what they were doing. Part of *our* job was to satisfy ourselves that they did.

"The rest of our job was to satisfy ourselves and everyone else in the picture that we could handle the machine once they gave it to us. Here was where our flying experience came in, and our ability to make fast, correct decisions in the clutch. We knew we had to practice and practice until we were sure that we could not only stay on top of any specific emergency that might come up on a flight, but that we could also handle troubles when they came in bunches."[31]

REFERENCES

1. Ken Kettlewell, *Our Town: New Concord, Ohio, The Birthplace of John Glenn*, Express Press, Lima, Ohio, 2001
2. *Ibid*
3. Frank Van Riper, *The Astronaut Who Would Be President*, Empire Books, New York, NY, 1983
4. Lt. Col. Philip N. Pierce and Karl Schuon, *John Glenn: Astronaut*, George G. Harrap & Co. Ltd., London, UK, 1962
5. The My Hero Project, *My Hero: Extraordinary People on the Heroes Who Inspire Them*, Chapter, *Senator John Glenn*, Free Press, New York, 2007
6. John Glenn with Nick Taylor, *John Glenn: A Memoir*, Bantam Books, New York, NY, 1999
7. *Teen Ink Online* magazine, unknown author. Website: *http://www.teenink.com/nonfiction/celebrity_ interviews/article/5438/John-Glenn-AstronautSenator*
8. *Ibid*
9. John Glenn with Nick Taylor, *John Glenn: A Memoir*, Bantam Books, New York, NY, 1999
10. Carpenter, S., Cooper, Jr. L, Glenn, Jr., J., Grissom, V., Schirra, Jr., W., Shepard, Jr., A., and Slayton, D., *We Seven*, Simon and Schuster Inc., New York, NY, 1962
11. *Teen Ink Online* magazine, unknown author. Website: *http://www.teenink.com/nonfiction/celebrity_ interviews/article/5438/John-Glenn-AstronautSenator*
12. Scott Montgomery and Timothy R. Gaffney, *Back In Orbit: John Glenn's Return to Space*, Longstreet Press, Inc., Atlanta, Georgia, 1998
13. John Glenn with Nick Taylor, *John Glenn: A Memoir*, Bantam Books, New York, NY, 1999
14. *Ibid*
15. *Ibid*
16. Lt. Col. Philip N. Pierce and Karl Schuon, *John Glenn: Astronaut*, George G. Harrap & Co. Ltd., London, UK, 1962
17. John Glenn with Nick Taylor, *John Glenn: A Memoir*, Bantam Books, New York, NY, 1999
18. Carpenter, S., Cooper, Jr. L, Glenn, Jr., J., Grissom, V., Schirra, Jr., W., Shepard, Jr., A., and Slayton, D., *We Seven*, Simon and Schuster Inc., New York, NY, 1962
19. *Ibid*
20. Frank Van Riper, *Glenn: The Astronaut Who Would be President*, Empire Books, New York, NY, 1983
21. *Ibid*
22. *Ibid*
23. John Glenn with Nick Taylor, *John Glenn: A Memoir*, Bantam Books, New York, NY, 1999
24. Dr. Stanley C. White email message to Colin Burgess, 14 April 2010
25. Dr. Robert B. Voas, interviewed by Summer Chick Bergen for NASA JSC Oral History series, Venice, VA, 19 May 2002
26. Jim Lovell and Jeffrey Kluger, *Lost Moon: The Perilous Voyage of Apollo 13*, Houghton Mifflin, Boston, MA, 1994
27. Dr. Robert B. Voas, interviewed by Summer Chick Bergen for NASA JSC Oral History series, Venice, VA, 19 May 2002
28. *Ibid*
29. Carpenter, S., Cooper, Jr. L, Glenn, Jr., J., Grissom, V., Schirra, Jr., W., Shepard, Jr., A., and Slayton, D., *We Seven*, Simon and Schuster Inc., New York, NY, 1962
30. *Ibid*
31. *Ibid*

4

Delays and more delays

In the wake of NASA's announcement that Lt. Col. John H. Glenn, Jr. would be the first American to orbit the Earth, with a tentative launch date set for 20 December 1961, an excited expectation began to ripple across the United States, with people eager to follow the progress of this historic event. His upcoming flight began to occupy an increasing number of columns in newspapers and magazines across the country. It was huge news in the making.

As it turned out, a good deal of patience would be needed as the launch date continually slipped for a variety of reasons – mostly weather related – but this just enhanced the growing interest in the flight. Both Glenn and his nation were ready to take that monumental leap into space flight history. NASA was fully aware of the significance of this orbital flight and the expectations of people, not only across America, but all around the world. Everything had to be perfect on a mission where a much-loved American hero was set to be launched on the first manned flight to be powered by an Atlas rocket, which had a checkered history, both of complete success and spectacular failure. The last thing NASA and the nation needed was to witness John Glenn perish in a massive fireball live on television, caused by undue haste to get him into orbit.

THE LONG WAIT BEGINS

The capsule for the MA-6 mission had arrived at Cape Canaveral on 27 August 1961, where it underwent a detailed system-by-system examination and tests to verify its configuration.

The examination, conducted in Hangar S, involved functional testing of the spacecraft systems to observe in detail the performance of them all, some of which had changed as a result of information obtained from the MA-5 orbital flight. Any discrepancies, no matter how trivial, were scrutinized for their significance. Glenn and his backup, Scott Carpenter, participated in all of these checkouts at the Cape and reviewed all design changes.

© Springer International Publishing Switzerland 2015

C. Burgess, *Friendship 7*, Springer Praxis Books, DOI 10.1007/978-3-319-15654-5_4

John Glenn training on the centrifuge at the Naval Air Development Center, Johnsville, Pennsylvania. (Photo: NASA)

This participation gave them an intimate familiarization with the spacecraft and a thorough understanding of its systems. The checkout was followed by launch complex operations in which the Atlas launch vehicle and Mercury spacecraft were mated and tested to ensure that the two were mechanically, electrically, and radio-frequency compatible.[1]

Early in December, Glenn and Carpenter moved into the astronauts' modest quarters on the second floor of Hangar S, where they would live ahead of the flight while working on the procedures trainer or in the spacecraft itself. Meanwhile the trickle of reporters turning up at the Cape had turned into a flood. But it would turn out to be a longer stay than anyone had anticipated.

Spacecraft No. 13, which Glenn would fly, inside Hangar S. (Photo: NASA)

NASA had hoped to launch the MA-6 mission on the Atlas launch vehicle soon after it arrived at Cape Canaveral on 30 November 1961, wanting to send an astronaut into orbit in the same calendar year as the Soviets. By early December, however, it was apparent that the mission hardware would not be ready for launch until early 1962, and it was not fair to keep everyone on the job, as recalled by Glenn:

"Finally, the Project Mercury brain trust decided that with Christmas approaching, rather than hold the whole operation – literally thousands of people, including engineers, technicians, the medical staff, ground control, teams at the tracking stations around the world, and more thousands who were watching and waiting, the press included – hostage to the whims of the weather and the mission's demanding technology, it was better and safer to stand down. I agreed with the decision. As much as I wanted to go, it was far better if everybody had their minds on the mission when the time came. NASA told everybody to go home for the holidays, and announced a date of January 16."[2]

Christmas came and went, and suddenly it was 1962. The Glenn family welcomed in the New Year by watching the Times Square crowd of revelers on their TV and listening to Guy Lombardo's orchestra play *Auld Lang Syne*. The following day, the Mercury spacecraft was mated to Atlas 109D in preparation for Glenn's flight.

John Glenn in Hangar S with his spacecraft in the background. (Photo: NASA)

CREATING AN INSIGNIA

Among other peripheral duties to which Glenn now had to attend was the naming of his spacecraft. Alan Shepard had dubbed his MR-3 capsule *Freedom 7*, and Gus Grissom followed suit by patriotically naming his *Liberty Bell 7*, which also alluded to the shape of the Mercury capsule. The numeral 7 was originally derived from the McDonnell production number allocated to Shepard's capsule, but because it also coincided with the number of Mercury astronauts they decided to continue the trend, particularly as it conveyed a sense of teamwork. Glenn had a few names in mind but he also wanted to involve his family in his flight, and during one visit home asked Lyn and David to come up with some suggestions. They eagerly accepted the challenge.

McDonnell engineers prepare to mate the spacecraft to the Atlas booster. (Photo: NASA)

"They pored over a thesaurus and wrote dozens of names in a notebook," Glenn later revealed in his published memoir. "Then they worked them down to several possibilities, names and words including *Columbia, endeavor, America, Magellan, we, hope, harmony,* and *kindness*. At the top of their list was their first choice: *friendship*. I was so proud of them. They had chosen perfectly."[3] McDonnell's Mercury capsule No. 13 thus became *Friendship 7.*

The names on Shepard's and Grissom's capsules had simply been spray-painted in white on the side using a pre-cut stencil, but Glenn wanted more than a plain block-letter name on his capsule. What he had in mind was something similar to the artwork that usually adorned the nose of American fighting aircraft. He felt that would be far more appropriate.

Following his request, he was put in touch with Cecelia ("Cece") Bibby, a graphic artist employed by the Chrysler Corporation, who did contract work for NASA. She happily drew for Glenn what she felt was an appropriate insignia, which he liked, and Bibby was hired to hand-paint the design on the side of the MA-6 capsule.

Cece Bibby painting the *Friendship 7* logo on the side of the spacecraft as John Glenn continues his training. (Photo: NASA)

John Glenn with the finished logo. (Photo: NASA)

"Glenn wanted *Friendship 7* done in script," Bibby revealed in a 2005 interview. "He wanted it applied by hand and not by stencil and a can of spray paint. The first trip out to the pad was to find the area where I was to do the painting on the spacecraft and to measure, because I [wanted] to do a large cartoon so I could trace it onto the capsule. I would use chalk to transfer my design to the capsule." But Cece Bibby encountered problems the very first time she entered the gantry's White Room. "When I got up to the top of the gantry I encountered the Pad Leader [Guenter Wendt], who informed me that women weren't allowed up there. I was told to leave immediately. I told him he'd have to take it up with John Glenn and I went ahead and did my job."

Glenn and the other Mercury astronauts were so impressed by Bibby's work that she was also commissioned to paint her original insignias on the sides of the capsules flown during the three final Mercury missions.[4]

The launch delays meant extra training time for Glenn, seen here going through flight rehearsals in the procedures trainer. (Photo: NASA)

A PROCESSION OF DELAYS

The first postponement of 1962 came when NASA announced that the launch date was being delayed a week until 23 January. This did not overly concern Glenn, as he had enough on his plate to keep him occupied, working on the procedures trainer through 70 simulated missions, and solving 200 simulated system failures. Every day of delay increased his confidence that he would be 100 percent ready to fly.

While Glenn and Carpenter continued to work as a closely knit team, they were members of the Mercury astronaut team that occasionally loved to play jokes or "gotchas" on

22 January 1962: Mercury capsule No. 13 mounted atop Atlas 109D on Launch Pad 14 with the egress facility extended to the capsule. (Photo: NASA/KSC)

each other to relieve some of the stress inherent in their training and other responsibilities. When asked about an oral history interview that he had conducted for NASA with Roy Neal in which he said that he could not think of any funny – or even tellable – stories from those days, Carpenter had laughed.

"It was a white lie of sorts, and even now I can't think of any except one that I played on John Glenn, when, in his little two- or four-cylinder Prinz – all the other guys had big muscle cars, but John has this little, tiny two-horsepower Prinz convertible – and he and I were in that car, late for an airplane flight out of Baltimore going somewhere, with the top down, and I was getting ready the tickets and all our baggage in the front of the car. And I got the tickets out and realized that if we lost those tickets, if the wind got ahold of them, we were really in trouble. But all the receipts and other paperwork in the ticket envelopes, I just let fly out of the car and go back on the freeway, and told John that, 'Oh gosh, there go our tickets!' We didn't have time to drive five miles for a place to turn around and go back for tickets. I told him, 'Well John, the tickets are gone. We just…' And he was disappointed, but he kept his cool. And then I told him, 'I've got the tickets – I was just joshin' you!' And he took it like you'd expect him to do, with equanimity."[5]

Among the many issues that cropped up during this time was one of Glenn asking to take photographs of the Earth while in orbit. But NASA was reluctant, believing this would distract him from carrying out far more important tasks. They also argued that no cameras then available were simple enough to operate on the ground, let alone be adapted for use in weightlessness. There would be a camera on board *Friendship 7*, but in his auto-biography Glenn says this was simply "an ultraviolet spectrograph with only a single roll of film for taking pictures of the Sun and stars."[6]

One day, Glenn happened to be in Cocoa Beach to get a haircut, and he dropped into the drugstore next door to purchase a few small items. His eye fell on a small camera for sale in a display case and he asked to see it. This was a Minolta Hi-Matic, and he was delighted to see that it featured automatic exposure, meaning that he would not have to fiddle with a light meter and f-stops any time he wanted to take a photo. He purchased the camera (it cost him $45), showed it to his NASA bosses, and finally convinced them that once modified, it would perform a useful function during his flight without detracting from other work. Eventually he received permission to carry it on his flight.

In order to overcome the problem of Glenn operating the camera while wearing his thick pressure suit gloves, NASA engineers turned the camera upside down, attaching a pistol grip and special buttons to control the shutter and film advance, and relocated the eyepiece to the bottom – now the top – of the camera so Glenn could target the constellation Orion for the spectrographic ultraviolet photography he was scheduled to perform. Once it was ready he tried operating a few cameras while wearing his pressure suit gloves, and found the Minolta was by far the easiest to use.

The camera that John Glenn would carry into orbit. (Photo: NASA)

Early on Monday, 22 January, technicians giving the capsule a final checkout discovered a fault in the all-important life-support system. One built-in safety measure in *Friendship 7*, as in all of the Mercury capsules, was that in the event of a sudden cabin depressurization automatic devices would instantly seal up Glenn's spacesuit and pipe all further compressed oxygen only to him, and none to the cabin. The fault was a valve that could have increased Glenn's oxygen flow so heavily that it would all be consumed in three hours. This would be disastrously short of the five hours total time of the planned three-orbit flight. NASA decided that two components of this closed-circuit breathing system needed to be replaced, a time-consuming job which forced the launch date shift from Wednesday to Saturday, 27th at the earliest.[7]

Glenn parked outside of the Cape's Mercury Control Center. (Photo: NASA)

CLOSE, BUT ANOTHER DELAY

Everything seemed set for the Saturday, 27 January launch, and Glenn was ready to go. He was woken by Dr. Douglas at 2:00 a.m., showered and shaved, and then sat down to a low-residue breakfast along with Bill Douglas, MSC Director Robert Gilruth, Associate Director Walt Williams, MSC Cape Operations Chief G. Merritt Preston, Deke Slayton and a surprise visitor for Glenn in the form of General David M. Shoup, Commandant of the U.S. Marine Corps.

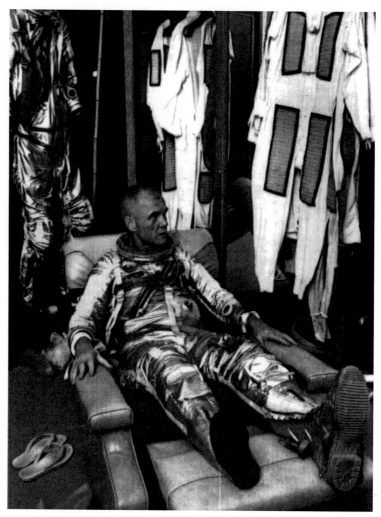

Suiting up for the 27 January launch attempt. (Photo: NASA)

Following breakfast, Douglas attached all the EKG and sensor leads to pre-selected spots on Glenn's body (marked by dot tattoos), before suit technician Joe Schmitt assisted the astronaut into his pressure suit. Glenn walked out of Hangar S at 4:46 a.m. carrying his small portable air-conditioning unit and boarded the transfer bus for the four-mile ride over to Launch Complex 14. Once there, he rode the gantry elevator up twelve stories to the canvas-covered White Room along with Bill Douglas and Deke Slayton.

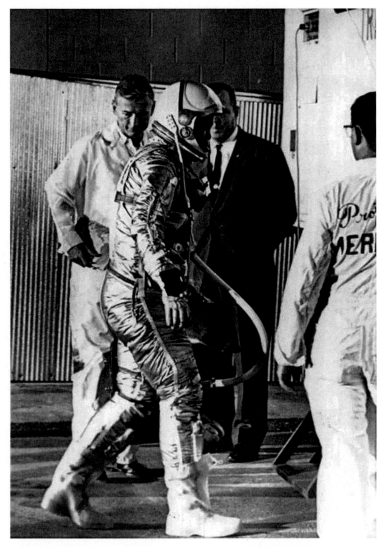

27 January 1962: At the start of a busy day, Glenn prepares to enter the transfer van which will carry him out to the launch gantry. Behind him in white overalls is suit technician Joe Schmitt. (Photo: Associated Press)

After Glenn had disposed of his anti-dust overshoes, Scott Carpenter and pad leader Guenter Wendt assisted him to slide feet-first into *Friendship 7*. Then Joe Schmitt buckled Glenn in and plugged his suit into the capsule's air supply.

Everyone knew that there was a potential problem with the weather. The giant tracking cameras at the Cape would have the important function during the vital first minutes of the flight of visually showing whether the Atlas injected *Friendship 7* into the proper low orbit. If the correct insertion into orbit was not achieved, their first-hand information would allow a quick descent from space off the coast of Africa, where cruising ships were waiting for just such an abort. As it was, a lingering blanket of cloud that had gathered during the night did not burn off with the rising Sun, and optimism was not running high as the hours ticked by.

"I lay there on my back in the contour couch for nearly six hours [officially recorded as five hours thirteen minutes], wishing the gantry would pull back, signaling a break in the clouds and imminent lift-off. But they didn't break, and the launch window closed at twelve-thirty in the afternoon."[8]

The countdown had reached T-13 minutes, but that was it. The hatch was opened and a perspiring Glenn was assisted out of the capsule. A new launch date was later decided; the first day of February.

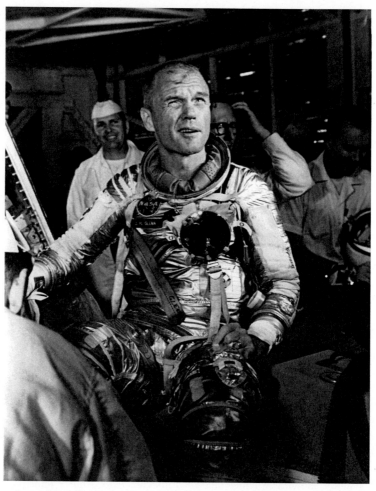

A hot and perspiring Glenn after being removed from *Friendship 7* on 27 January. (Photo: NASA)

A disappointed astronaut climbs into an automobile for the ride back to Hangar S. (Photo: NASA)

In announcing the revised launch date, NASA press spokesman Col. John Powers said there were no specific problems in either the Atlas rocket or the capsule when the decision was made to postpone the flight. "When we got to 9:30 a.m., which meant only a two-orbit flight, we thought it was not worth the effort to proceed. Weather was the imminent factor, but it got that way because of technical holds earlier in the countdown."[9]

At Palm Beach, Florida, President Kennedy had watched the prelaunch preparations on television. He was notified two minutes in advance of the formal announcement at the Cape. Although disappointed, the President praised the space agency for showing "good judgment" in postponing the shot.[10]

An estimated 10,000 people had spent the previous night on the Cape beaches. Many had bloodshot eyes and sand in their hair from overnight parties around campfires.

NASA press spokesman Col. John ("Shorty") Powers. (Photo: NASA)

That number had swollen by another 65,000 who had joined them early in the morning, watching the Sun come up at 7:13 a.m. Most of the spectators were armed with cameras, binoculars and the occasional transistor radio. The hitherto happy crowd had been sprawled for seven miles along the beachfront, sizzling steaks and sausages over charcoal broilers or eating hot dogs, but following the announcement of the postponement a collective groan went up. Beach tents came down, picnic lunches were canceled, and soon thereafter the traffic heading away from the Cape was in gridlock. Disappointment was rife. Some people said they would return for the next launch attempt, but others were resigned to leaving Florida either to return home or to resume work. Some of the luckier people had rooms booked at Cocoa Beach's two dozen motels, which had all 1,500 rooms sold out all week.

One incident that did come out of that day, which featured prominently in the movie of Tom Wolfe's seminal astronaut book *The Right Stuff*, involved Vice President Lyndon Johnson and Annie Glenn. Those close to Annie knew that she had suffered badly from childhood with a pronounced stammer, although this was never mentioned in the contemporary press. That morning the Vice President had turned up at the Glenn home at 3683 North Harrison Street, Arlington, Virginia, after the delay had been announced, accompanied by a number of favored reporters and photographers. In fact hordes of reporters, photographers, and assorted voyeurs were already present, at the ready in case a family member came outside. In *The Right Stuff*, Tom Wolfe described the whole scene as chaotic. "The lawn, or what was left of it, looked like Nut City," wrote Wolfe. "There were three or four mobile units from television networks with cables running through the grass. It looked as if Arlington had been invaded by giant toasters." When Johnson sent his aides through the crowd into the house, they returned shortly after to say that Mrs. Glenn had refused them – and Johnson – entry. As John Kennedy's liaison with the space program, the Vice President was furious at being slighted in this way.

Loudon Wainwright from *Life* magazine was in the Glenn household with a photographer under the agreed and exclusive contract to cover the private lives of the astronauts and their families, and he remembered the confrontation well.

"Wolfe has got it kind of right and kind of wrong," he said. "I was in the house with the photographer and we didn't want Johnson to come into the house because, if he came into the house, all of the other Washington press would come in with him. Annie didn't want him to come in, either. She felt it was a real invasion of her privacy to have about fifty people in the house. She really didn't need the Vice President, but she didn't want to be rude to him. As I remember it, she held him off. It was with my complete agreement, because I didn't really want him in there either."[11]

When asked about the incident, Annie responded, "It had come to the point where John was almost going to be launched and it had to be scrubbed. I was really tired and I was getting a migraine. I was just plain going to bed, and then there was all of that commotion and everything."[12]

Weary, hot and strained after spending nearly six hours in his spacecraft, Glenn then had to support his wife in not allowing Vice President Lyndon Johnson into their home. (Photo: Associated Press)

John Glenn also recalls the problems that morning. "That was the day of the big scrub. I had been on top of the booster for five and a half or six hours, flat on my back. When I got out, I was sweaty. I was tired. And I got out of the spacesuit and was in a terry cloth robe on the way to the showers when someone in NASA said they wanted to see me. Instead of saying: 'Buck up, we've just scrubbed the Free World's first effort at manned orbital flight,' the question was whether I was going to let the Vice President into my house. I just told them to go with whatever Annie wanted, and that whatever agreement we had made before still stood .... I just thought that wasn't the agreement we had made [with *Life*]. And Annie was up there under the gun. I went off to the showers, and when I came back the Vice President was gone."[13]

Following the cancellation of the 27 January launch at T-13 minutes, it was decided to replace the carbon dioxide absorber unit because it had approached the end of its service life. The peroxide system also had to be drained and flushed to prevent corrosion, and the pyrotechnics were disconnected and shorted as a safety measure. It only took a single day to accomplish these tasks.[14]

COPING WITH POSTPONEMENTS

On 30 January, to the ongoing annoyance of many – especially the world's media who were stationed around the Cape – NASA decided that the manned MA-6 launch would be further postponed until 13 February due to "technical difficulties with the launching booster." The problem, the nature of which was reported to the press on 1 February, was the discovery of a leak in the inner bulkhead of the Atlas rocket's fuel tank during pre-flight preparations. After fueling operations, which involved loading the Atlas with the kerosene-like RP-1 fuel, an inspection had revealed residual fuel contaminating insulation material in the upper portion of the fuel tank. No other elements of either the Atlas or the Mercury spacecraft were affected by the leak. It was estimated that the leak would take 4–6 days to be repaired. The repairing of the leak, followed by the necessary retesting and launch preparation and other operational considerations, had dictated rescheduling the launch to the announced date.[15]

Col. Powers quoted Glenn as saying: "Sure, I'm disappointed, but this is a complicated business. I don't think we should fly until all elements of the mission are ready. When we have completed all our tests satisfactorily then we'll go."[16]

Following the announcement of an almost two-week delay in the launch, hordes of news media representatives packed their bags and left Cocoa Beach for their homes and offices. Most would return prior to the new launch date.

Interviewed at his home on 3 February, Glenn said that the rescheduled MA-6 launch on 13 February "can only bode for success." Surveying the crowd of edgy newsmen parked on his lawn, he remarked that "it looks like Hangar S was not such a bad place after all." Two days later he visited the White House at the invitation of the President.

John Glenn with German rocket designer Wernher von Braun. (Photo: NASA)

On 6 February, NASA further announced that the MA-6 launch attempt had been slipped by a day to February 14. Repair and test work on the Atlas had not been completed. During this delay, all six flight batteries and the parachutes were replaced.

At a regular White House press conference, President Kennedy was asked about the eighth postponement to Glenn's flight. He replied, "Well, it is unfortunate. I know it strains Colonel Glenn. It has delayed our program. It puts burdens on all those who must take these decisions as to whether the mission should go or not. But I think we ought to stick with the present group who are making the judgment and they are hopeful still of having this flight take place in the next few days. And I'm going to follow their judgment in the matter even though we've had bad luck."[17]

It came as no surprise to those inured to the ongoing launch postponements when, on 14 February, NASA decided to call off its ninth scheduled attempt to launch the MA-6 mission because a weather briefing held at midnight warned of storms continuing to lash the Atlantic landing area. As the ships in the recovery fleet would be able to remain at sea for less than a week, it was decided to proceed on a day-to-day basis, starting on the 15th, in an attempt to make quick use of any break in the weather.

After 57 days of waiting since the first scheduled attempt on 20 December, Glenn was "taking it very well," according to Project Mercury psychologist, Robert Voas. He told reporters that, "There is no evidence he is building up any frustrations or annoyance. He is anxious to get in three full orbits of the Earth and would prefer to wait rather than not have everything right for the full plan." This followed rumors within the press corps that Glenn was becoming so upset with the delays that there were plans to replace him with

Carpenter, but this was rejected as totally untrue.[18] These rumors had their origins in a statement by Dr. Constantine Generales, Jr., a space medical researcher with no connection to NASA, who suggested that Glenn should be taken off the mission after so many delays. "Anxiety had undoubtedly built up in Glenn's subconscious from all the stops he has encountered," Generales had declared.[19]

Dr. Bill Douglas strongly dismissed any suggestions that Glenn was displaying any signs of tension that might affect his performance, or that he should be replaced on the flight. "I'm as close to this man as I am to my brother," Douglas emphasized, "and I couldn't let my brother fly if I thought he would be in danger. If I detected anything wrong, I would take immediate action." He reiterated that Glenn had full confidence in the program and, as an experienced test pilot, was used to such delays.[20]

READY TO TRY AGAIN

On February 16, following an early morning weather briefing, Mercury Operations Director Walt Williams advised that weather conditions once again precluded a launch attempt of the MA-6 mission, and Tuesday, 20 February was announced as the earliest possible launch date. Public Affairs spokesman Col. Powers said there was "a fifty-fifty" chance the flight would go on that date. Meanwhile it was rumored at the Cape that a one-month postponement might have to be ordered if weather or technical difficulties prevented a launch that week. Notified of the decision at 12:50 a.m., a resigned Glenn said, "I guess it was to be expected. We all knew the weather was marginal."

NASA Flight Director Chris Kraft nailed the frustration that was now being felt by close to 1,000 reporters from newspapers, radio and magazines, both American and from overseas, who were at the Cape for the launch. "The reporters' moods were as foul as the weather," he wrote in his 2001 memoir. "They were stuck in Cocoa Beach, overrunning their expense accounts, and the only story they could write was about one more delay."[21]

Four days later, everyone was ready to try again. All they needed was a clear sky, calm seas, and a spacecraft and booster that would pass every test for launch.

In a 2002 interview with the author, Scott Carpenter said he and Glenn both remained philosophical about the many postponements. "The delays didn't bother either of us much. It gave us both more time to get ready. And you figure it's just part of the job – if something fails, you wait for it to be fixed and try again."[22]

John Glenn was more than ready. The day before the revised launch date he assured Operations Director Walt Williams at a meeting that he was okay, "But there was a local storm front moving across Canaveral at the time," he later recalled, "and that night the weather people gave us only a fifty-fifty chance that it would clear in time for a launch the following morning. These odds had never taken us very far on earlier tries, so I was not exactly hopeful. It seemed more likely that we might have to wait another day."

At this point, Glenn's task was to get a good night's sleep, so he retired to bed around seven o'clock that evening.

One person who recalls that pre-launch day and another small problem with the Atlas booster is Terry Terhune.

Members of the press and electronic media were becoming extremely edgy following numerous launch delays, impatient for a launch. (Photos: NASA)

"I was employed by Convair under a USAF contract to test and launch Atlas ICBMs.
I was a young propulsion engineer working on CX 12 (Launch Complex 12) when tasked
to accompany my lead engineer to CX 14 on second shift to monitor the Atlas propulsion
system of John Glenn's boost vehicle, to be launched the next day. The CX launch crew
had been working demanding hours and were resting for the next day's launch.

"The Atlas had been tanked with RP-1 fuel per procedure and all systems were secure.
Our only significant task was to monitor the sustainer engine that had demonstrated a
small fuel leak past its main valve. This was accomplished via a procedure that required

the removal of a drain screw off the engine bell cooling tubes. A very small amount was allowed. We checked the quantity hourly and it was determined to be an acceptable seep. We later received a glossy photo of the lift-off for our efforts.

"Fast forward many years. John was now a Senator and attending a fundraising event at the Astronaut Hall of Fame in Titusville, Florida. I was then working for NASA and had the photo proudly displayed in my office – but with no signature. I removed the photo from the frame and it was presented to John with the aforementioned story. He gave his shy Boy Scout smile and with tongue in cheek said, 'A leak? Well now, if I had known, I might not have flown.' He then signed the photo."[23]

REFERENCES

1. G. Merritt Preston: "Spacecraft Preparation and Checkout," from *Results of the First United States Manned Orbital Space Flight*, February 20, 1962, NASA Manned Spacecraft Center
2. John Glenn with Nick Taylor, *John Glenn: A Memoir*, Bantam Books, New York, NY, 1999
3. *Ibid*
4. Lawrence McGlynn article, "Breaking through the Glass Gantry" for *collectSPACE*, 7 August 2005, online at: *http://www.collectspace.com/news/news-080705a.html*
5. Scott Carpenter telephone interview with Colin Burgess, 18 December 2002
6. John Glenn with Nick Taylor, *John Glenn: A Memoir*, Bantam Books, New York, NY, 1999
7. Article, "Glenn's Flight: The Full Report," *Space World: The Magazine of Space News (Special Issue)*, Vol. 2, No. 5, April 1962, pg. 2
8. John Glenn with Nick Taylor, *John Glenn: A Memoir*, Bantam Books, New York, NY, 1999
9. *Sydney Morning Herald* newspaper, "U.S. Manned Space Bid Delayed at Least for Three Days," issue 29 Jan 1962, pg. 1
10. *Ibid*
11. Frank Van Riper, *Glenn: The Astronaut Who Would Be President*, Empire Books, New York, NY, 1983
12. *Ibid*
13. *Ibid*
14. G. Merritt Preston: "Spacecraft Preparation and Checkout," from *Results of the First United States Manned Orbital Space Flight*, February 20, 1962, NASA Manned Spacecraft Center
15. *Ibid*
16. NASA *Space News Roundup*, NASA Manned Spacecraft Center, issue 7 Feb. 1962, pg 1
17. John F. Kennedy Presidential Library and Museum, *President Kennedy's News Conference, 14 February 1962*. Online at: *http://www.jfklibrary.org/Research/Research-Aids/Ready-Reference/Press-Conferences/News-Conference-24.aspx*
18. *Sydney Morning Herald* newspaper, "U.S. Calls Off Ninth Space Shot Attempt," issue 16 February 1962, Pg. 1
19. Frank Van Riper, *Glenn: The Astronaut Who Would Be President*, Empire Books, New York, NY, 1983
20. *Ocala Star-Banner* (Florida) newspaper, uncredited article "Tuesday Earliest for Orbit Flight, Glenn Still Choice," 16 February 1962, pg. 1
21. Chris Kraft and James Schefter, *Flight: My Life in Mission Control*, Penguin Putnam Inc., New York, NY, 2001
22. Scott Carpenter telephone interview with Colin Burgess, 18 December 2002
23. Terry Terhune email correspondence with Colin Burgess, 8 December 2014

5

"Godspeed, John Glenn"

"As the weeks went by from December and into February, it must have begun to look as if I were never going to go," John Glenn would later write about the frustrating series of launch delays. "We would have been ready at almost any time, but first bad weather, then troubles with the gear, then more bad weather kept piling up the delays. We were disappointed, of course, each time we had to stop. Scott and I were most concerned, however, about the 20,000 people who were directly involved with the program – the launching crews, the men in the recovery fleet, the technicians standing by at the tracking stations scattered around the world and all of the Mercury group who were on hand at Cape Canaveral. Each delay meant that these crews had to recycle their work, go through the long checklists and start up the countdown all over again. I think people normally build up to a peak when they are preparing for an event as complicated as this, and here we had a situation where we kept building up psychologically and nothing happened. It was like crying 'Wolf!' over and over."[1]

WORKING AS A TEAM

Despite the numerous delays, the crews who were the objects of Glenn's concern kept on steadfastly working toward the goal of getting *Friendship 7* off the launch pad – whenever that might be. Meanwhile Scott Carpenter was always there to back up and encourage his friend. He and Glenn had become good friends during the astronaut selection process, and this friendship had evolved into a durable working relationship and a mutual determination to make MA-6 the success it deserved to be.

However, as he told the author in 2002, Carpenter often found it difficult to keep up with his friend's passion, drive and dedication to the flight.

"You know, he never quit, and I could stay with him in all of the long hours and everything else we had to do, but he would not relax – he was just single-minded. And that's one of his virtues. The only thing that really wore me out was his passion for running, and I don't care for running that much. There are other exercises that I'd prefer to just pounding the pavement. But he did wear me out, because of his dedication. The fact that he got the

© Springer International Publishing Switzerland 2015

C. Burgess, *Friendship 7*, Springer Praxis Books, DOI 10.1007/978-3-319-15654-5_5

Scott Carpenter and John Glenn working together as a team at the Cape. (Photo: NASA)

attention didn't bother me at all; the fact that there were delays didn't bother either of us much; it gave us both more time to get ready. And you just figure it's part of the job – if something fails you just wait for it to be fixed and try again."[2]

The two men remained stoic in the face of the delays; they kept on studying charts, reviewing the flight plan and working in the trainers to practice in-flight procedures, repeatedly practicing maneuvering the capsule with the manual controls.

"As it happened," Glenn later stated, "this extra homework was very useful."

MISSION CONTROL: A MOST IMPORTANT TEAM

Gene Kranz enjoys an incredible history. He was part of America's manned space flight program from the outset, and as a flight director in NASA's Mission Control he actively participated in the Mercury program through to the final Apollo lunar landing mission before moving on to the space shuttle program. In a talk in 2000 he spoke about these pioneering days, and how some incredibly adept teamwork brought about the safe return of John Glenn and other astronauts from potentially catastrophic situations. The teams at Mission Control, Kranz noted, were ready for almost any contingency.

"The tools that we used, to contain the risks of our business are leadership, trust, values and teamwork. Leadership; capable of focusing on the objective and then building, no matter what, to get there. Trust; which allows us to make every second count and find every option. Values; which build the chemistry that we need to be successful. Teamwork; which literally assures our victory.

"When we started in 1960, our world was vastly different. In that decade, we would have three major political assassinations. Our country was torn by the beginning of the conflict in Vietnam. The civil rights movement was just starting within the nation. The Cold War with Russia guided every aspect of our foreign policy. That was basically our rationale for the American space program. Computers existed only in laboratories. There were no real global communications. During that decade, American students would riot on our campuses. Then, in 1961, a young, brash, energetic President called John F. Kennedy gave us a dream. He said, 'We choose to go the Moon, in this decade, and do other things,

Gene Kranz (lower left) with flight directors Glynn Lunney, John Hodge, and Chris Kraft. (Photo: NASA)

not because they are easy – but because they are hard.' He issued this challenge as we were struggling to put an American in Earth orbit.

"So engineers, people experienced in flight test, and a small group of Canadians and Englishmen joined with the Mercury Seven astronauts at Langley Field in Virginia. Our goal was simply to place an American in space. Our boss was Walt Williams, the toughest man I have ever known – a brawler, more fitted to working as a longshoreman than leading the American space program! But in the business of aircraft flight test, Williams was a legend. He was the pioneer director of the high speed test station that later became known as Edwards Air Force Base. He was also the project manager and engineer for the X-1 rocket ship, that in 1947 took Chuck Yeager and the world into the age of supersonic flight. Williams' deputy was appropriately named Christopher Columbus Kraft. How would you like to work for a guy with a name like Christopher Columbus? Chris was a pioneer of Mission Control. He launched each one of the Mercury missions, and he was the mentor, teacher, tutor for the first generation of people who became known as Mission Control. Mercury Control did not have a single computer. There were only three mainframe computers supporting all of Project Mercury. Two of these mainframes were at Goddard Space Flight Center in Greenbelt, Maryland. They were the first fully transistorized computers developed by IBM. At the island of Bermuda we put a reliable vacuum tube model, and we always launched eastward over the island, so that if we lost communications during powered flight we had a team in place that would tell the crew what to do – when the engines shut down, if they were go or no-go, whether they were in Earth orbit or not.

"Worldwide communications consisted of a 60-word-per-minute teletype network, so with only a couple of weeks training, and proficiency in Morse Code we sent young men to thirteen tracking stations around the world. They went into the heart of Africa, to the island of Zanzibar, to Australia, on islands in the Pacific and ships at sea. These were sites literally at the end of the Earth. The risks to these young people were very high, as the sixties was the end of the European colonial era. Much of Africa was in a state of revolution or civil war. The controllers at Kano (Nigeria) were twice rescued by the army, from rioting crowds. In Zanzibar they would go to work under the protection of the Queen's Royal Rifles, or the Gordon Highland Regiment. When their work was done at the tracking station, they would put a gunner at the top of the stairs. Our ships seldom went into ports in South America, because of concerns about mining and sinking. Normally they would anchor offshore. These young people at the tracking station were our eyes, ears and voice as the spacecraft passed overhead. They were given incredible responsibilities, and they grew to be the leaders for Project Apollo.

"We first ventured into space in Project Mercury, and each one of those launches proved to be a chapter in the history books of space flight. Our mistakes were violent, they were brutal; they were visible. Two of our first three Atlas rockets exploded shortly after launch. When we tried to launch our first Redstone rocket, we sent the firing command, the engine ignited, it lifted four inches off the pad – and then the engine shut down. Through some miracle, it landed right back on the launcher. The escape tower fired – went to an altitude of about four thousand feet, and then came hurtling down towards the viewing stand. The booster engineers in Mercury Control were speaking in German. We did not understand what they were saying, and literally did not know what to do.

"We were fortunate in those days, as our nation understood that there was no achievement without risk. There certainly weren't any guarantees in this new business that was called space flight. We put six Americans in space in the first two years. There were close calls in virtually every one of those missions, especially John Glenn's and Scott Carpenter's – but we got every one of our crew members back. By the time we had finished Project Mercury, we had learned that man could live in space, but we had learned much more. We had learned about leadership. Leaders must have integrity; they must say what they mean, then go out and execute it. Leaders must accept risk, but they must never pass uncertainty to their people. We also learned a lot about ourselves. Most of us came in from flight test, and our egos were big. It was really tough at times to get people to work together. But we found that success can only come as a team, and so we became a team.

"This is the story of Mission Control. It's a story of people tied together by leadership, values, teamwork and trust. People who accept risk, make decisions. People who move beyond bowing to the conventional and who consistently drive for this thing that is so difficult to achieve, that is called perfection. These are the standards that we live by – the foundations of Mission Control. These are the standards that are represented by every person who ever works in our business, from the very top level down to the newest rookie coming in."[3]

BEGINNING THE BIG DAY

It was Tuesday, 20 February, and John Glenn's historic day began early when Bill Douglas snapped on the lights in the astronaut's Hangar S bedroom at 2:20 a.m. The astronaut was in the top bed of a double-decker bunk in the room. "Okay, Marine, rise and shine," Douglas said. "Today's the day." Like everyone else, he was hopeful that this would indeed be the day *Friendship 7* finally took to the skies. The constant delays were getting on everyone's nerves, and he was understandably concerned about the effect they might be having on the astronaut.

"Good morning, Bill; how's the weather look?" Glenn enquired, as he stretched out and prepared to get up. He had already been awake for about half an hour.

"Socked in right now, John," the astronauts' physicians cautiously responded. "But aerology says there's a good chance it'll clear up before launch time."

As he jumped down from the bunk and headed off to take a shower, Glenn still seemed cautiously optimistic after all the delays. "Better keep your fingers crossed, Bill," he called back over his shoulder.

According to the Mercury Control Center (MCC) procedures log, at 3:40 a.m. the flight control team noted that they were "up and at it." They began to conduct a radar check, and even though ionospheric conditions were poor, controllers believed the weather situation would improve. They continued to check booster telemetry and MCC's voice intercom system, both of which were found to be in good order. Soon after, they traced a faulty communication link that was meant to be obtaining information about the capsule's oxygen system. Within minutes they had corrected the problem.[4]

At 4:27 a.m. Christopher Kraft, sitting at his flight director's console, received word that the global tracking network had been checked out and was ready.[5]

Mercury Control Center prepares for the MA-6 mission. (Photo: NASA)

Glenn enjoys breakfast with Bill Douglas, Merritt Preston, and Deke Slayton. (Photo: Associated Press)

After a shower and shave, now comfortably attired in a terry-cloth robe, Glenn sat down to breakfast with Bill Douglas, Deke Slayton, Walt Williams, and Merritt Preston, chief of pre-flight operations. He ate two eggs, a small filet mignon steak, toast with jelly, orange juice and a cup of Postum, a caffeine-free coffee substitute. Following this, the by-now familiar pre-launch routine began once again. First, Douglas and his medical team

Bill Douglas gives John Glenn a complete medical checkup prior to the flight. (Photo: NASA)

carried out a full medical examination; he then attached bio-sensors to Glenn's body to record and transmit his vital signs during the flight. Meanwhile, Scott Carpenter had traveled over to the launch pad to monitor the capsule preparations and run final systems checks.

At 4:30 a.m., as he chatted with Deke Slayton, Glenn donned his long underwear and was then inserted into his 20-pound silver pressure suit with the help of NASA's suit technician Joe Schmitt. As he later wrote, Schmitt was an old hand at these suit-ups. "Joe is one of the most conscientious and hard-working men I know. He has presided over all of our suitings-up, and he is the kind of dedicated citizen and skillful craftsman who makes Project Mercury go."[6]

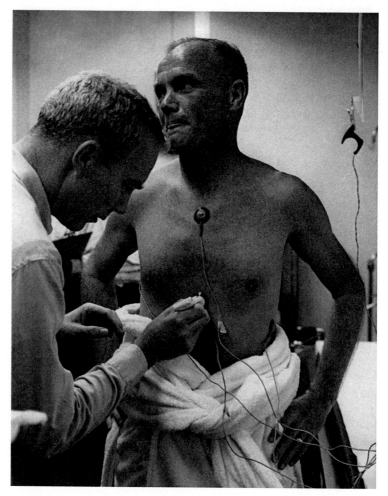

Attaching sensors to pre-determined spots on the astronaut's body. (Photo: NASA)

When the suit, gloves, boots and other accessories were on and secured, Schmitt conducted a suit pressure test along with Bill Douglas, who ran a main air supply hose from the suit into a small fish tank, to check the purity of the air in the suit. In his 1999 memoir, Glenn related that Douglas had told him if any of the fish died it meant that the suit air was bad. A short while later he stunned Douglas by casually mentioning that a couple of the fish were floating belly-up. Shocked, Dr. Douglas rushed over to the tank before realizing that Glenn was pulling his leg. Shortly after, Glenn put on his white helmet, which was sealed in place, and then he walked out of Hangar S carrying his portable air conditioning unit.

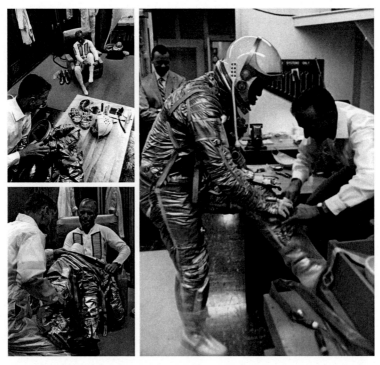

NASA suit technician Joe Schmitt prepares to assist Glenn in donning his space suit, finishing up with the protective overshoes. (Photos: NASA)

Glenn recalls there were about 150 people gathered outside Hangar S as he emerged and made his way to the transfer van. "This was a pretty big crowd for five in the morning," he reflected.[7] There was a smattering of applause and he waved back before climbing the steps into the van for the four-mile journey out to Launch Pad 14. Bill Douglas and Joe Schmitt boarded the van with Glenn, along with meteorologist Ken Nagler. Deke Slayton had already made his way out to the pad while Glenn was being suited up. The weather still looked about a fifty-fifty proposition, but was said to be clearing.

With dawn approaching, the van arrived at the launch pad at 5:17 a.m., some 20 minutes behind schedule. Glenn remained optimistic, "and as we pulled up to the pad we opened the blinds on the van windows, and I could see the gantry. It was a beautiful sight. It was still dark, but the big arc lights shone white on the Atlas booster. It looked like something out of another world."[8]

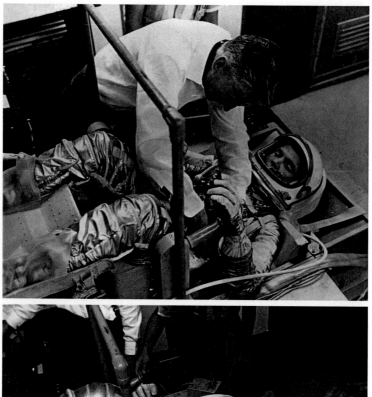

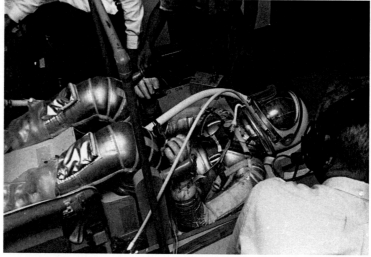

Joe Schmitt conducts a suit pressurization test. (Photos: NASA)

Accompanied by Bill Douglas, Glenn departs Hangar S for the transfer van that will transport him out to the launch pad. (Photo: NASA)

He did not leave the van; in fact everyone remained inside for 45 minutes during two launch holds. The first one was a planned hold to check on weather conditions at the Cape, and then the blockhouse called for a technical hold in order to replace a misreading radar-tracking transponder in the Atlas rocket's guidance system. Along with Deke Slayton, who had now joined him in the van, Glenn kept on patiently reviewing the flight plan and weather updates. Meanwhile the capsule and booster validation checks were progressing normally. Then it was time and Glenn carefully stepped down from the van, followed by his entourage. The weather was still far from ideal, but everyone remained cautiously hopeful.

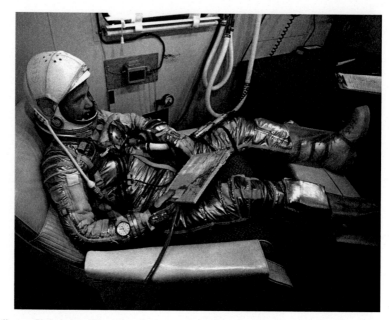

Holding a flight map, Glenn heads out to the launch pad aboard the transfer van. (Photo: NASA)

"Finally, at one minute before six, we all walked to the elevator which was waiting for us at the foot of the gantry. The crews clapped and waved their good wishes. I nodded my thanks, and felt very good as I realized how much they were with me. Everyone is a team-mate with you on a morning like this. Then I stepped into the elevator and rode up to the eleventh deck – spacecraft level."[9]

READY TO GO

Stepping out of the elevator into a canvas-clad, environmentally controlled chamber known by all as the White Room, Glenn could see a hive of activity going on around *Friendship 7*, with technicians making last-minute checks. Official photographers were busy recording the scene, and everyone seemed to feel that this was going to be the big day. A smiling Scott Carpenter strolled over and said that all the capsule checks had been finished and everything looked good. He had provided tremendous support as Glenn's backup pilot. "Scott is much more to me, however, than a willing col-league," Glenn would later observe. "He is also a close friend, and that morning we both felt emotions which could not be expressed when we shook hands and he wished me good luck."[10]

Glenn removes the outer protective shoes prior to his insertion into *Friendship 7*. (Photo: NASA)

Glenn then had a quick friendly banter with McDonnell engineer and pad leader Guenter Wendt as he prepared for his insertion into the waiting capsule with the count-down holding at T-120 minutes to the scheduled launch time. It was a routine they had practiced many times over the months. First of all he disposed of the outer protective boots that he had worn for the journey from Hangar S. Then, with the assistance of a couple of technicians, he carefully eased himself feet-first through the hatch and squeezed into the cramped interior. It was 6:03 a.m., just seven minutes after sunrise, and word was passed through to the Mercury Control Center that the astronaut was now settled inside *Friendship 7* and being strapped in.

Once he was comfortably seated, Glenn noticed a problem with the respiration sen-sor, a thermistor attached to the astronaut's microphone in the air stream of his breath. The sensor had shifted from where it had been fixed during an earlier simulated flight. NASA aerospace doctor Stanley White pointed out to Flight Operations Director Walt Williams that a correction could only be made by opening the suit, a very tricky opera-tion atop the gantry. Eventually they decided to disregard the slipped thermistor, even though faulty data would result. White advised the range to ignore all respiratory transmissions.[11]

Joe Schmitt now continued his duties by leaning in over Glenn and loosely fitting all the restraint straps. He then assisted in plugging the suit into the environmental control system (ECS) of the capsule. They had gone through this procedure many times, but were prepared for any problem to crop up. So when the support on his left-hand helmet micro-phone broke and had to be replaced, causing another hold in the countdown, contingency

Glenn is inserted feet-first into the cramped confines of the spacecraft. (Photos: NASA)

plans meant that there was a backup helmet in the transfer van. This was quickly sent up to the hatch level where, despite having very little room to work in, Schmitt managed to replace the damaged helmet without the need for Glenn to egress the capsule.

The countdown was resumed at 6:25 a.m. Time was now of the essence, because if the mission was to go for the planned three orbits it had to launch by 9:30 at the latest to allow for a daylight recovery at sea. The only options beyond that were to scrub the launch yet again, or reduce the flight to just two orbits.

Once Glenn had been fully secured by Schmitt and all the systems checks made, it was time to close the hatch on the capsule. An hour had passed since he had clambered inside *Friendship 7*, and he was eager to go. The only real problem now was the weather.

Glenn would describe the hatch closure procedure as "an interesting moment … when things really begin to come home to you. Up until now, people have been reaching into the capsule with their hands to fix something behind you, pat you on the shoulder or shake your hand. You know how strongly they all feel, and you feel strongly, too. But there is no real feeling yet of being on your own. Then, suddenly, there are no more hands. The hatch is closed. It is quite a moment – and a good one."[12]

It was now 6:59 a.m. As he had practiced over and over again, Glenn systematically ran through his check list, looking for – but hoping he would not find – any faults or potential problems. Below him, he was aware of the Atlas rocket creaking and fuming, but they were familiar sounds and he was able to ignore them as he ran his eyes over his instruments.

Cloud masses continued to hover over the launch area, and many of the weary newsmen were prepared for the fact that the launch might not proceed that day. Then, a little after 7 a.m., one of the Cape weather men, Harlan G. Higgins, noticed that the wind was shifting to drive the clouds away and that the temperature was becoming cooler. He quickly phoned Ernest Amman, the weather support man in Mercury Control, and told him that the chances for launch now looked promising.

Then, at 7:25 a.m., just five minutes before the original lift-off time, there was a further delay when technicians discovered a broken bolt on the capsule hatch. The hatch had to be reopened and the bolt replaced, which caused another hold in the countdown.

Pad leader Guenter Wendt recalled the broken bolt incident with a humorous anecdote. On that day James McDonnell, the founder of McDonnell Aircraft (affectionately known to his employees as "Mr. Mac"), was watching the launch preparations on a television set in the Mercury Control Center. The picture he was watching came from a fixed black-and-white camera located in the White Room that showed the hatch of the Mercury spacecraft.

"After the astronaut had entered the capsule I got permission from the test conductor to install the explosive hatch," Wendt stated. "The work was performed by McDonnell technicians dressed in white coveralls with big McDonnell logos on their backs. After the cabin hatch was in place one of the pre-drilled attachment bolts broke during the final torque application. I informed the test conductor and he requested a 'hold' in the launch count. A conference was held in the control center and I was told to open the hatch and to replace the platenut inside the cabin that held the broken-off piece of the bolt with a new platenut. After a 22-minute delay the platenut was replaced and the hatch was reinstalled. But for the whole time the TV that the world saw was the backs of our technicians with the big McDonnell logo on it. Mr Mac (who was very tight with money) was extremely delighted to get the 22 minutes of TV advertising without having to pay a dime for it." Once the bolt had been replaced the hatch was closed up once again and the cabin purge was started.[13]

A check of the cabin oxygen leakage rate indicated 500 cubic centimeters per minute, which was well within the design specifications. Meanwhile the sky began to lighten in the east amid clearing clouds.

Guenter Wendt was the last person to have physical contact with John Glenn before the hatch was closed for a second time. "I saw a man who was getting impatient. That launch had been postponed ten times before it went off. He was anxious to have it go this time. That was the look in his eyes. It was, 'Let me go!'"

Glenn gives the OK signal for hatch closure. (Photo: NASA)

Years later, Guenter Wendt proudly shows off a fine painting by space artist Ed Hengeveld (commissioned by Noah Bradley) which depicts the "broken bolt" incident on Pad 14. Inset: The actual broken bolt as preserved in lucite by Wendt, plus the associated platenut. (Photos courtesy of Joseph Hiura; reproduction permission granted by Ed Hengeveld)

FINAL PREPARATIONS

There was a very positive sign for everyone just after 8:00 a.m. when claxons suddenly blared outside of the White Room, a clamorous but encouraging warning to the launch personnel to remove themselves from the gantry area.

As part of his launch duties, Scott Carpenter had made his way across to the block-house, a few hundred yards from the launch pad. It was protected by walls of concrete and steel that were ten feet thick. It had no windows; the launch crew inside used periscopes to view the launch pad.

When the countdown reached a point where he knew that his friend did not have any scheduled duties, Carpenter patched a call through to the capsule from Annie Glenn at the Glenn home in Arlington. The family was watching the launch preparations on television, very much aware of the noisy crowd of reporters and others who were patrolling outside, hoping for a glimpse of the famous family. The situation had not been helped by many newspapers printing their home address. Despite her anxiety over this and her husband's impending launch, Annie was able to talk on the private line with Glenn as the shrouding rust-red gantry was rolled back from the pad. Now with a clearer view, Glenn told his wife about the improving weather conditions over the Cape and all the activity going on ahead of the launch. Although he could not see them, he knew many thousands of people were lining nearby beaches and roadways, some of whom were sufficiently dedicated to have been going through that same routine for several months and through all the launch delays. His fingers were figuratively crossed for them.

The assembled news media were weary but hopeful that this day would be launch day. (Photo: NASA)

The scene inside the blockhouse as launch preparations continued. (Photo: NASA)

Once again crowds of spectators had assembled at every vantage point for the launch. (Photo: NASA)

After talking with Annie, Glenn also spoke briefly with Lyn and David, before his wife came back on for a few more seconds. With a little hesitation, he asked Annie if some audio tapes he had prepared earlier had arrived at the house. On these tapes he had recorded two private and very personal messages – one meant for Annie and one for the children – which he wanted them to listen to if anything happened to him during the flight. Annie confirmed that the tapes had arrived. He closed out the conversation with a phrase he had often used to reassure her whenever he went overseas or on a lengthy assignment, which somehow made it all seem a little easier: "Hey, honey, don't be scared. Remember, I'm just going down to the corner store to get a pack of gum." She gave her customary response, "Don't be long." He then said he would talk to her again after he landed that afternoon. After each of them had closed with "I love you," Glenn had to hang up.

After he had spoken to Annie, Glenn felt more than ready for the adventure ahead. As he would write some years later in his memoir:

Friendship 7 atop the Atlas booster and ready for lift-off. (Photo: NASA)

"In a mirror near the capsule window, I could see the blockhouse and back across the Cape. The periscope gave me a view out over the Atlantic. It was turning into a fine day. I felt a little bit like the way I had felt going into combat. There you are, ready to go; you know all the procedures, and there's nothing left to do but just do it. People have always asked if I was afraid. I wasn't. Constructive apprehension is more like it. I was keyed up and alert to everything that was going on, and I had full knowledge of the situation – the best antidote to fear. Besides, this was the fourth time I had suited up, and I still had trouble believing I would actually take off."[14]

In his post-flight pilot's report, he would also discuss the final few minutes of the countdown. "The initial unusual experience of the mission is that of being on top of the Atlas launch vehicle after the gantry has been pulled back .... If you move back and forth in the couch, you can feel the entire vehicle moving very slightly. When the engines are gimbaled, you can feel the vibration. When the tank is filled with liquid oxygen, the spacecraft vibrates and shudders as the metal skin flexes. Through the window and periscope the white plume of the lox [liquid oxygen] venting is visible."[15]

WINDING DOWN THE COUNTDOWN

By this time the weather was deemed acceptable for launch and was no longer an issue. At 8:57 a.m. an announcement came from the Mercury Control Center that all systems were in a "go" condition. Listening on their car radios or tiny portable radios, thousands of spectators taking every vantage point surrounding the Cape were cheered by the news. In homes across the country, 100 million people were tuned in on their television sets, eagerly watching the grainy black-and-white images emanating from Cape Canaveral.

For the next 30 minutes, telephones stopped ringing across the nation. In New York's Grand Central Station thousands of commuters, normally in a rush to get somewhere, stood still and silent as they watched live coverage of the launch pad activities on huge television screens. Inside the White House, President Kennedy interrupted breakfast with some Democratic congressional leaders so they could gather around a television set to watch the launch.

At the Glenn home in Maryland, his family and a small gathering of close friends were watching the launch preparations on a row of three television sets, each tuned to one of three major networks. As the launch time drew nearer, Annie was nervously rubbing a necklace her husband had given her when they were dating. Meanwhile their 16-year-old son David was busy mapping out *Friendship 7*'s planned flight path on the family globe. His 14-year-old sister Carolyn was trying not to show her concern by concentrating on shining a pair of shoes.

Elsewhere, Glenn's parents were also watching the launch preparations from his boyhood home in New Concord. "I'll bet he's cool as a cucumber," his father mentioned to a reporter, and he was quite correct. Glenn's pulse rate at that time was registering between 60 and 80 beats per minute, which indicated he had no real apprehension or even dread about what was about to happen. In fact he was displaying far less stress than his family and loved ones, and those who were closely associated with his impending lift-off.[16]

A short hold occurred at T-22 minutes, when a valve stuck in the liquid oxygen system; but it was a relatively minor malfunction and was quickly corrected. Up to this time, it had not been much fun for the tens of thousands of people standing on the beaches near Cape

General Dynamics engineer Tom O'Malley (left) and NASA spacecraft test conductor Paul Donnelly pose with John Glenn during an earlier flight test. (Photo: NASA)

Canaveral. Many had been at the Cape since mid-January and had organized little "trailer towns," complete with their own "mayor." Mission announcer, Col. John Powers, popularly known as "the voice of Mercury Control," had been at his post since 5 o'clock that morning, and he went on the air to advise the waiting public of the status of the countdown and the cause for the present hold.

Then, six and a half minutes before lift-off, a power failure occurred in the Bermuda tracking station which meant that *Friendship 7* must wait until power was restored. The space authorities were going to take as few risks as possible. Fortunately the power was quickly restored and the hold only lasted for two minutes. Then Glenn heard the ground controllers report in, as directed by the General Dynamics test conductor in the blockhouse, Irish-American aerospace engineer Tom O'Malley. He was the one who would press the black button to initiate the firing sequence that would send the Atlas on its way. All the responses to his queries came in fast and positive.

"Communications?"

"*Go!*"

"ASCS?"

"*Go!*"

"Aeromed?"

"*Go!*"

Voices tumbled over each other as the clock ticked down. It all sounded quite chaotic, but in fact this was a well-rehearsed and orderly procedure.

"*Minus forty!*"

"Status check: pressurization?"

"*Go!*"

"LOX tanking?"

"I have a blinking, high-level light."

"You are *Go!*"

At T-40 seconds, in accordance with procedures, Glenn automatically placed his left hand on the abort handle. The voices continued.

"Range operations?"

"*All are clear to launch!*"

"Mercury capsule?"

Glenn gave an emphatic "*Go!*"

"All prestart panel lights are correct. The ready light is on. Eject Mercury umbilical. All evacuate!"

"*Mercury umbilical clear.*"

Up to this time the umbilical cord had been providing external power and cooling for *Friendship 7*. Watching through his periscope, Glenn saw it fall away with what he later described as a loud plop. Next, the periscope retracted automatically and his outside view was gone.

"All recorders to fast: T-18 seconds and counting. Engine start!"

O'Malley's boss, Byron MacNabb, General Dynamics' Astronautics Base Manager, was seated in MCC and he radioed a quiet Irish prayer directed to O'Malley: "May the wee ones be with you, Thomas."

Immediately after came O'Malley's response, "The good Lord ride all the way."

In the blockhouse, Tom O'Malley prepares to push the button that will initiate the Atlas firing sequence. (Photo: NASA)

The crowded scene at Grand Central Station, New York, as the launch begins. (Photo: NASA)

Then, also from the blockhouse, came the gentle voice of Scott Carpenter as he intoned three of the most memorable words in the entire Mercury program: "Godspeed, John Glenn."

As he later explained, it was an entirely spontaneous remark:

"The two previous Mercury flights were powered by the Redstone, a tiny rocket that couldn't provide enough power to give John the speed he required for orbital flight. What he needed, and what everyone hoped the Atlas would provide, was speed. I had not pondered this, just as Neil Armstrong maintains he'd never pondered the phrase, 'One small step for [a] man, one giant leap for mankind.' It just popped out of thin air. He needed speed, his name was John Glenn, and it was sort of a salute to a friend, and a plea to the higher power. Godspeed."[17]

Glenn actually heard none of the prayerful transmissions or Carpenter's message. He would listen to them post-flight on a recording of the transmissions. What he was hearing, and concentrating on, was the voice of Alan Shepard continuing his steady countdown:

"…eight, seven, six, five, four, three, two, one, zero!"

"*Ignition!*"

"*Lift-off!*"

It was 9:47:39 a.m.

As Glenn would later describe that moment: "I could feel the engines light off as the capsule vibrated from their ignition, and I could hear a faint roar inside the capsule. The booster stood fast on the pad for two or three seconds while the engines built up to their proper thrust. Then the big hold-down clamps dropped away and I could feel us start to go. I had always thought from watching Atlas launches that it would seem slow and a little sluggish, like an elevator rising. I was wrong; it was not like that at all. It was a solid and exhilarating surge of up and away. Al Shepard received a signal that I was lifting off and confirmed it for me over the radio. The capsule clock started right on time and I reported this. 'The clock is operating,' I said. 'We're under way.'"[18]

REFERENCES

1. Carpenter, S., Cooper, Jr. L, Glenn, Jr., J., Grissom, V., Schirra, Jr., W., Shepard, Jr., A., and Slayton, D., *We Seven*, Simon and Schuster Inc., New York, NY, 1962
2. Scott Carpenter telephone interview with Colin Burgess, 18 December 2002
3. Gene Kranz, email correspondence with Colin Burgess, 8–18 November 2014, plus Apollo 13 talk on C-Span, recorded 12 November 2000. Available at: *http://www.c-span.org/video/?160396-5/book-discussion-failure-option-mission-control*
4. *Postlaunch Memorandum Report for MA-6; Procedures Log*, Mercury Control Center, 20 February 1962
5. *Ibid*; Kranz memo; memo, Stanley C. White to Kraft, "Summary Report on Test 5460 (MA-6)," 22 February 1962
6. Carpenter, S., Cooper, Jr. L, Glenn, Jr., J., Grissom, V., Schirra, Jr., W., Shepard, Jr., A., and Slayton, D., *We Seven (Chapter: "The Mission")*, Simon and Schuster Inc., New York, NY, 1962
7. *Ibid*
8. *Ibid*
9. *Ibid*
10. *Ibid*
11. Stanley C. White memo, *Postlaunch Memorandum Report for MA-6; Procedures Log*, Mercury Control Center, 20 February 1962
12. Carpenter, S., Cooper, Jr. L, Glenn, Jr., J., Grissom, V., Schirra, Jr., W., Shepard, Jr., A., and Slayton, D., *We Seven (Chapter: "The Mission")*, Simon and Schuster Inc., New York, NY, 1962
13. Guenter Wendt memo, 25 July 2003, courtesy of Joseph Hiura
14. John Glenn with Nick Taylor, *John Glenn: A Memoir*, Bantam Books, New York, NY, 1999, pg. 258
15. John H. Glenn, Jr., *Pilot's Flight Report: Results of the First United States Manned Orbital Space Flight, February 20, 1962*, NASA Manned Spacecraft Center publication, pp. 119–136
16. Jeff Lyttle, "John Glenn: An American story," *Columbus Monthly* magazine, Columbus, OH, issue August 1998
17. Scott Carpenter telephone interview with Colin Burgess, 18 December 2002
18. Carpenter, S., Cooper, Jr. L, Glenn, Jr., J., Grissom, V., Schirra, Jr., W., Shepard, Jr., A., and Slayton, D., *We Seven (Chapter: "The Mission")*, Simon and Schuster Inc., New York, NY, 1962

6

A drama-filled mission

Generating 367,000 pounds of thrust, the 125-ton Atlas 109D tore away from the launch pad with a brutal roar that swept across the Cape. As programmed, the 42-inch umbilical cord that carried electrical connections to the base of the rocket fell free, and John Glenn's last physical connection with the Earth was gone. Majestically, but ever so ponderously in the eyes of the onlookers, the Atlas rose from Launch Pad 14. Those thousands of people watching across the Cape area collectively held their breath. The lift-off was also seen by what (to that time) was the largest television audience in history, with an estimated 135 million viewers watching live pictures coming from the Cape. Others were listening to radio broadcasts of the event which were carried all around the world, including the Soviet Union. If America had wanted a big show, this was certainly it.

BLAZING A TRAIL

"The fiery plume and cloudy trail were immense as the rocket cleared the tower," was the recollection of pad leader Guenter Wendt. "Fly! Fly! We craned our heads, looking more steeply into the bright sky as the silver bird soared higher. No one could hear it over the sound of the rocket engines, but every mouth seemed to be shouting and the fists were triumphantly raised into the air.

"The Atlas was gulping fuel down at the rate of 2,000 pounds a second. With each passing second it gained speed. And with each second it became lighter allowing it to gain even more speed."[1]

Air Force Lt. Dee O'Hara was the astronauts' nurse, and as the time to launch had drawn near she was standing nervously at the gate to the forward medical area where she had set up a mobile trauma unit, her eyes firmly fixed on Launch Pad 14. As the time to launch wound down she nervously fiddled with her sunglasses, first taking them off, twirling them, then putting them back on. In the background she could hear the droning of the five helicopters which would go to Glenn's rescue in the event of a launch emergency and carry him straight to the medical area for immediate first aid. Then, if necessary, he could be flown to a hospital, depending on his injuries.

© Springer International Publishing Switzerland 2015
C. Burgess, *Friendship 7*, Springer Praxis Books, DOI 10.1007/978-3-319-15654-5_6

After weeks of delays and frustration, *Friendship 7* finally takes to the skies. (Photo: NASA)

"We had communications there," O'Hara recalled. "As well as myself there was a security guard, my med tech, and maybe a few other NASA people."

As the rocket rose into the sky ("It was kind of heart-stopping when it did get off") Lt. O'Hara shaded her eyes with her hands and watched with her heart in her throat until it disappeared from sight. Her shoulders slumped in relief. Soon after, she turned to the security guard and asked to borrow a dime from him for a very welcome cup of coffee.[2]

As the mighty Atlas surged ever higher into the morning sky over the Cape, John Glenn prepared himself for the phase of the flight known in short terms as Max Q – the point of the ascent in which aerodynamic pressure on the rocket and its human cargo was at its greatest. This would occur at around 35,000 feet. As Guenter Wendt explained, "It was the one point where the odds were greatest that the vehicle might explode. I crossed my fingers in my pocket."[3]

Spectators lining nearby beaches cheer as the Atlas rocket leaves the launch pad. (Photo: NASA)

Newsreel and newspaper reporters, photographers, and other media representatives watch as John Glenn's historic journey begins. (Photo: NASA)

Fellow Mercury astronaut Deke Slayton had seated himself right beside CapCom Alan Shepard in the Mercury Control Center at the Cape, listening intently as the Atlas rolled onto its intended northeast bearing. Glenn reported that this maneuver was going well, and then the Atlas began to rattle a shaky path through Max Q. "A little bumpy along about here," he radioed back, his voice vibrating a little as the shaking grew in intensity. So far everything about the flight had gone as predicted. After about 30 seconds the vibrations

Dee O'Hara in discussion with John Glenn. (Photo: NASA)

CapCom Alan Shepard, Deke Slayton, and Dr. Bill Douglas watch the progress of the flight from the Mercury Control Center. (Photo: NASA)

began to dampen out and 1 minute and 16 seconds after lift-off Shepard reported that the rocket had now passed through Max Q. Glenn replied that the flight was "smoothing out real fine," although the g forces were now building and he found himself straining against a load on his body of 7 g's. However this was known and anticipated and would soon pass.

A little over two minutes into the flight, the two main engines shut down right on schedule in a phase known as BECO, or booster engine cut-off. By this time the Atlas was 40 miles high and 45 miles from the Cape.

"BECO! … BECO!" Glenn said, almost shouting. "I saw the smoke go by the window."

"It was a little easier ride for John than it had been for Al or Gus – seven g's as opposed to eleven," Slayton noted in his memoir. "We got booster cut-off at 2 [minutes] 9 [seconds] into the mission; the Atlas pressed forward on its single sustainer engine …. The guidance system began to pitch the vehicle forward so that it would go into orbit horizontally. The Atlas kept wiggling around as the guidance system tried to keep the single sustainer engine firing through the center of gravity. John later said he felt like he was on the tip of a spring-board. As the fuel was expended, gas was pumped into the Atlas propellant tanks to keep them from collapsing. That only made things louder."[4]

As Glenn later recalled: "There was no sensation of speed, however, because there was nothing outside to look at as a reference point. The only time I took my left hand off the abort handle during this period was to be ready to jettison the escape tower at exactly 2 minutes and 34 seconds after lift-off in case it didn't leave automatically. That 900 pounds of extra weight would just waste fuel I needed for getting into orbit."[5]

He need not have worried; the tower atop of *Friendship 7* automatically rocketed away in what Glenn later described as "a momentary cloud of flame and smoke." He felt a slight bump as this happened, and watched through his window as the tower rapidly disappeared from sight. The dynamic load was appreciably less now, at around 1.5 g's. Things were looking good.

"Roger, reading you loud and clear, *Seven*," Shepard said as Glenn reported that g forces were building up once again. "Cape is 'go.' We're standing by for you."

"Roger … Cape is 'go' and I am 'go'," was Glenn's response. "Capsule is in good shape. Cabin pressure holding steady at 5.8 [pounds per square inch], amps is 26. All systems are 'go'."

The next crucial event would occur when the sustainer engine cut-off (SECO) took place. "Five minutes into the flight, if all went well, I would achieve orbital speed, hit zero-g, and if the angle of ascent was right, be inserted into orbit at a height of about a hundred miles," Glenn later recorded in his memoir. "The sustainer and vernier engines would cut off, the capsule-to-rocket clamp would release, and the posigrade rockets would fire to separate *Friendship 7* from the Atlas."[6]

The Atlas had propelled *Friendship 7* out of the atmosphere by this time, and built up sufficient speed that all that was needed was a final push from the sustainer engine to drive the spacecraft into orbit.

At 5 minutes and 4 seconds into the ascent, the sustainer engines cut out. "SECO!" Glenn reported. Then, "Posigrades fired okay." The firing of these small rockets served to push the capsule away from the Atlas. The capsule's periscope extended, and *Friendship 7* went into an automatic rotation so that the blunt end would be facing the direction of travel, allowing Glenn to look back out through the window. Moments later the excitement in his voice was easily evident when he reached a memorable and historic moment in his flight.

"Zero-g and I feel fine. Capsule is turning around. Oh, that view *is tremendous!*"

"Roger," Shepard confirmed. "Turnaround has started."

Illustration showing the separation of *Friendship 7* from the Atlas booster. (Photo: NASA)

Glenn's voice remained clear but filled with the obvious wonderment of what he was seeing and experiencing as he spoke again.

"The capsule is turning around and I can see the booster doing turnarounds just a couple of hundred yards behind me. It was beautiful."

"Roger, *Seven*," Shepard said. "You have a 'go' for at least seven orbits."

"Roger, understand go for at least seven orbits," Glenn responded.

This burst of communication has often been misinterpreted by commentators over the years, who mistakenly took this to mean that Glenn's original flight plan was for a seven-orbit mission. It was not.

"Shepard meant that the Cape computers indicated the insertion of the capsule was good enough for a minimum of seven orbits," Glenn later explained. "It probably would have been good enough for 17 or 70 if we had been able to carry enough fuel and oxygen for such a mission."[7]

INTO ORBIT

Now traveling smoothly in Earth orbit at 17,500 miles an hour and experiencing the pleasant effects of zero-g, Glenn loosened his chest strap and started work on his busy flight schedule. The first orbit was essentially one of familiarizing himself with his new environment, as well as establishing contact and passing data down to ground stations as they passed below. He gave the controls – both manual and automatic – a thorough check as he swept across the Atlantic, and everything was exactly as it should be. With this established to his satisfaction, he began the pleasant task of viewing and reporting on what he could see through his window.

As he passed over the Canary Islands shortly after launch, Glenn captured this image of cloud cover over the North Atlantic. (Photo: NASA)

The Atlas Mountains as photographed from *Friendship 7*. (Photo: NASA)

"I spotted the Canary Islands about 15 minutes after lift-off and picked up the African coast a couple of minutes later. The Atlas Mountains [in Morocco] were clear through the window. Back inland I could see huge dust storms blowing across the desert as well as great clouds of smoke from brush fires along the edge of the desert. I reported the dust storms and the tracking station in Kano, Nigeria told me they had been going on for a week."[8]

Over the tracking station on Zanzibar, off the east coast of Africa, information began flowing into the station about Glenn's blood pressure, and it was time for him to work on a small exercise machine for the benefit of aeromedics who were interested in knowing how the human body might be affected by weightlessness. It was a simple device, basically a handle attached to an elastic cable. In operation, Glenn pulled the handle up to his chin using both hands, then allowed it to spring back between his knees. He had to do this once per second for thirty seconds while the bio-sensors attached to his body transmitted data via telemetry back to the ground. Meanwhile, the beauty of the Earth still astounded the astronaut.

"The first sunset, which came over the Indian Ocean, was a beautiful display of vivid colors – oranges, yellows, purples of all graduations – that extended out through the atmosphere for about 60 degrees on either side of the Sun. The Sun itself was so bright that I had to use filters to look directly at it. When the Sun was still high in the sky its light was more bluish-white than yellow. When it came through the window, it was similar in color and intensity to the huge arc lights we use at the Cape."[9]

During his first orbit, Glenn photographed storm clouds over the Indian Ocean, darkening as the Sun approached the horizon. (Photo: NASA)

Glenn was also mildly surprised that he was able to see some stars by day against the black sky, and when he observed them while in the Earth's shadow he was able to roughly calculate his position by referencing familiar features such as the constellation of Orion and the Pleiades star cluster.

One of his strongest memories was seeing the stars set some 18 times faster than when standing stationary on Earth. He was also able to observe the cloud cover over the planet, the extent and percentage of which surprised him. The Moon, which was nearly full, was out each time he crossed the Pacific, and he transmitted that the clouds "showed up crisp and clear in its light." He also reported seeing lightning effects in some clouds located to the north of his course.

As he began closing in on the west coast of Australia, Glenn came into radio contact with Gordon Cooper, who was acting as CapCom at the remote tracking station at Muchea, some 30 miles northwest of Perth. After a little friendly banter, Glenn passed on information about his visual sightings, the onboard systems, and his general condition.

Cooper then suggested that Glenn keep an eye out to his right for a pattern of lights on the ground. Meanwhile the astronaut was maintaining a high level of excitement at what he was experiencing.

"That was sure a short day," he observed.

"Say again, *Friendship Seven*," Cooper requested.

"That was about the shortest day I've ever run into," came the response.

"Kinda passes rapidly, huh?"

John Glenn was filmed throughout the flight to record his physical reactions and eye movements for better placement of instruments on later flights. (Photos: NASA)

After further discussions on Glenn's observations of star patterns and constellations, he was asked to give a blood pressure reading. Everything was still normal and going very smoothly. Then he reported seeing some lights below.

"Is it just off to your right there?" asked Cooper.

"That's affirmative. Just to my right I can see a big pattern of lights apparently right on the coast. I can see the outline of a town and a very bright light just to the south of it."

"Perth and Rockingham you're seeing there," Cooper advised.

"Roger. The lights show up very well and thank everybody for turning them on, will you?"

"We sure will, John."

FLYING INTO "FIREFLIES"

As Glenn passed over the next tracking station at Woomera in South Australia, they reported that his blood pressure was quite normal – under the circumstances – at 126 over 90. He confirmed that he felt fine, had no problems with his vision, and was free of any problems of nausea or vertigo from the head movements that he deliberately undertook occasionally.

Over the coral atoll tracking station of Canton Island in the South Pacific Ocean, roughly halfway between Hawaii and Fiji, he lifted the visor on his helmet and took his first meal – squeezing some applesauce into his mouth from a toothpaste-like tube. He reported that weightlessness had no affect at all on his ability to eat and swallow food.

Soon after this, he began to notice something unexpected outside his window that startled Glenn, and would be widely reported after his flight – the mysterious so-called "fireflies" no one could adequately explain at the time.

"The strangest sight of all came with the very first ray of sunrise as I was crossing the Pacific toward the U.S. I was checking the instrument panel and when I looked back out the window I thought for a minute that I must have tumbled upside-down and was looking up at a new field of stars. I checked my instruments to make sure I was right-side-up. Then I looked again. There, spread out as far as I could see, were literally thousands of tiny luminous objects that glowed in the black sky like fireflies. I was riding slowly through them, and the sensation was like walking backwards through a pasture where someone had waved a wand and made all the fireflies stop right where they were and glow steadily. They were greenish in color, and they appeared to be about six to ten feet apart. I seemed to be passing through them at a speed of from three to five miles an hour. They were all around me, and those nearest the capsule would occasionally move across the window as if I had slightly interrupted their flow."[10]

Glenn thought that they had planned for, and foreseen, everything that might happen on his flight, but this was something entirely out of the ordinary. He reported this mysterious phenomena to the next tracking station at Guaymas in Mexico, but even though controllers there acknowledged the unusual sighting he later said they were far more interested in giving him the retro-sequence time – the exact moment that the retrorockets would have to fire if he were to make a premature re-entry after a single orbit.

Ninety minutes into the flight, Glenn was in voice contact with Wally Schirra, stationed at the Point Arguello station in California. As he began to report on the spacecraft systems to Schirra, the first real inkling of a problem caught Glenn's immediate attention.

"I had just picked [Wally Schirra] up and was looking for a sight of land beneath the clouds when the capsule drifted out of yaw limits about twenty degrees to the right. One of the large thrusters kicked it back. It swung to the left until it triggered the opposite large thruster, which brought it back to the right again. I went to fly-by-wire and oriented the capsule manually."[11]

For a time, as he traveled east into brighter sunshine, Glenn watched as the mysterious "fireflies" rapidly diminished in number. He then tried switching back to automatic attitude control, but the problem recurred as the spacecraft once again began to swing to the right at around one degree per second. He switched back to manual mode and reported this to Alan Shepard at the Cape as he flew by. Mercury Control concurred with him that he should remain on fly-by-wire, which was working quite smoothly.

With this seeming to be the only major technical concern, *Friendship 7* slipped into a second orbit of the Earth.

ORBIT TWO AND SEGMENT 51

As he crossed over the Canary Islands on his second orbit, Glenn was beginning to feel a little uncomfortable because the temperature in his suit was higher than it should have been. To add to his woes, the stations located in Kano and Zanzibar noticed a worrying 12 percent drop in the spacecraft's secondary oxygen supply. As well, and due to problems with the control system, he had already canceled a number of tests, although none were considered crucial. "I had to cancel several of the experiments and observations which I wanted to make, including a series of tests on the Sun's corona, some measurements of the brightness of the clouds, a second meal – a tube of mashed-up roast beef – and some further tests of a pilot's ability in space to adapt himself to darkness. I was also unable to take as many pictures as I had intended."[12]

Having received the Cape's recommendation to remain on fly-by-wire, Glenn ran through conditions in the spacecraft cabin, and stressed that the only truly unusual aspect of the flight so far, apart from the trouble with the ASCS, were the small particles he had observed. He had been told that the President might be speaking with him by way of a radio hook-up, but this did not eventuate.

As the flight progressed, Glenn confirmed to himself that the seemingly ever-present "fireflies" were not caused by gas emanating from the reaction control jets, so they still remained a mystery. He also practiced swinging the spacecraft around in yaw, so that for a time he was facing in the direction of flight. The set maneuver was completed without any problem.

During much of the second orbit, Glenn worked on the control system, attempting to pin down a pattern of errors so that he could determine what was wrong and make allowances for it. "I could hear and feel the large thrusters outside of the capsule as they popped off their bursts of hydrogen peroxide, first in one direction and then in the other. I could feel the slight throb of the smaller nozzles when I cut them in."[13]

As he passed the two-hour mark of his flight, with the spacecraft back in its blunt-end-first attitude, he was experiencing his second sunset. He was still frustrated by continuing control problems, which he reported to the Zanzibar station.

"This *Friendship Seven*. My status: I am on ASCS, it is not holding all the time. My trouble in yaw has reversed. During the first part of the flight, when I had trouble over the west coast of the United States, I had a problem with the yaw, with no low thrust to the left; now I have thrust in that direction but do not have low thrust to the right. When the capsule drifts out in that area, it hits high thrust and drops into orientation mode, temporarily."

Zanzibar acknowledged the transmission, and then asked Glenn to report at length on his physical condition, switch settings, and conditions within his suit and the spacecraft.

Meanwhile, back at the Cape, technician Bill Saunders in Mercury Control was routinely scanning the bank of 90 meters in front of him, each of which provided some information on conditions aboard *Friendship 7*. Suddenly, everything changed. He had noticed that meter number 51 was flashing. It was a potentially serious anomaly, so he immediately informed flight director Chris Kraft and CapCom Alan Shepard, saying, "I've got a valid signal on Segment 51!"

Segment 51 was a reference to *Friendship 7*'s heat shield. The telemetry that Saunders had noticed indicated that the ablative fiberglass heat shield had somehow come loose – a dangerous development. By design, the shield was not firmly fixed to the blunt end of the spacecraft because it served a dual purpose; it was to be an insulator on re-entry and a shock absorber on splashdown. At a certain point after re-entry, it was supposed to drop loose and hang beneath the spacecraft as the bottom of an air-filled landing bag. With a blunt-end re-entry through the atmosphere, the Mercury capsule was designed to withstand temperatures of around 3,000 degrees Fahrenheit. Without its heat shield, however, a returning spacecraft would disintegrate into white-hot chunks and the astronaut would be incinerated. It was as simple, and as hazardous, as that.

A reproduction of *Friendship 7* in orbit, showing the blunt-end heat shield and the retro-package attached to the rear of the spacecraft. (Image generated with Celestia; 3D model created by James R. Bassett)

Manfred ("Dutch") von Ehrenfried had joined NASA in 1961, becoming a flight controller in the Flight Control Operations Branch. Legendary Flight Director Gene Kranz recalls that prior to the MA-6 mission von Ehrenfried "was a new recruit who joined us as a procedures officer in time for the Glenn mission. He had been teaching high school physics when President Kennedy set the lunar goal and was itching for a piece of the action."[14]

At first von Ehrenfried felt thrilled and privileged to be involved in this most historic mission, but then the team under Flight Director Chris Kraft found themselves dealing with the potentially calamitous Segment 51 situation. Not only was anxiety growing that John Glenn and his spacecraft might be in deep trouble, but the team had precious little time to come up with any sort of workable, lifesaving decisions. As he recalls:

"During John Glenn's flight, the famous 'Segment 51' signal indicated the landing bag had deployed. Since this doesn't happen in orbit there was great concern. If it was just loose how would that affect the heat shield and the aerodynamics of re-entry? Had the heat shield come loose? Was it just a faulty signal?

"Needless to say, a lot of communications went around the control center 'loops' trying to determine the true status of the spacecraft. What were the remote sites seeing? Was the signal just from one station?"[15]

First of all, Kraft had to check that the meter reading was not faulty, and asked Saunders for the signal strength on meter 51. A few moments later he had some truly bad news; the signal strength was 80 percent, and he knew the MA-6 mission was in deep trouble. If the reading had been zero, that would only indicate a bad meter. A reading of 35–40 percent

NASA Flight Director Chris Kraft in obviously less stressful times. (Photo: NASA)

meant it was probably just a poor switch connection, while a reading of 100 percent would indicate a grounded circuit. But Kraft knew from experience that an 80 percent reading could likely mean that the heat shield had indeed slipped.

Kraft made urgent contact with the other tracking stations to see if they were reading the same indication. Soon afterwards they all reported in, and all with the same response; they had received the same signal. If the signal was correct, as it seemed to be, then the mission was likely going to end in disaster and the death of John Glenn. It was a terrifying prospect, and von Ehrenfried said that immediate action was needed.

"After getting inputs from the flight controllers, Kraft began discussing the situation with Max Faget, the spacecraft designer, and Walt Williams, the Mission Director. There was grave concern. It was up to 'Procedures' to interface with the sites, so Gene [Kranz] got together with McDonnell engineers Ed Niemann and John Yardley to determine what we should ask the remote site flight controllers as the spacecraft went over their sites.

"There is a small room off the main control center floor that housed the teletype equipment operated by Andy Anderson. About five of us were in this room drafting the messages to all the sites asking them to report what they saw on telemetry and to ask Glenn questions. Some were really stressed and it showed in their faces.

"This was the first manned orbital mission. How would it end?"

Kranz was all too aware of the implications. "If the telemetry indication being reported was correct and the heat shield had come loose in orbit, John Glenn would have no protection from the 3,000 degree Fahrenheit re-entry temperatures. The capsule would become a meteor that flashed for but a few brief seconds during re-entry before burning up."[16]

By the time Glenn had reached the Indian Ocean a second time, he was aware that something had gone badly awry with the ASCS indicators. What he could see out of his window simply did not reconcile with the attitudes of yaw, pitch and roll being indicated on his instrument panel. When the CapCom on the Indian Ocean tracking ship, the *Coastal Sentry Quebec*, asked if he could see any constellations at this time, Glenn said he was too busy checking his instrumentation to notice. It was then that he got his first verbal indication that something might be amiss with the capsule's heat shield.

"We have [a] message from MCC," was the message relayed to him, "for you to keep your landing bag switch in off position. Over."

Glenn acknowledged the request, which he found a little puzzling at the time, and the mystery deepened even more seven minutes later when he was in contact with Muchea in Western Australia.

"Will you confirm the landing bag switch is in the off position?" asked Gordon Cooper.

"That is affirmative," Glenn replied. "Landing bag switch is in the center off position."

"You haven't had any banging noises or anything of this type at higher rates [of angular motion]?" Cooper's questions caused Glenn to wonder what was going on, as this was the second time he had been asked the same thing. He had not heard any unusual noises and, as far as he was aware, everything was essentially shipshape with his spacecraft.

"Negative," he responded.

"I had heard nothing," he later wrote. "Still, there was room for concern … the heat shield is made up of a thick coating of resinous material which is designed to dissipate most of the heat and energy picked up during re-entry and get it out of the capsule's system by melting and boiling away very slowly. This was the only thing that stood between me and disaster as we came through the atmosphere. If it was not tightly in place, we could be in real trouble."[17]

As he pondered this situation he was kept busy, checking on two warning lights which were not associated with the heat shield problem. One showed there was an excess of water in the cabin's environmental control system, and the other indicated that more fuel had been consumed than should have been the case in the automatic control system. However neither was a critical problem.

Then, as *Friendship 7* passed over Canton Island, the CapCom came online just as Glenn was stowing his camera after taking some further photos of the mysterious fireflies. With contact established, the usual questions followed about spacecraft systems. This occupied much of their transmission time before moving on to questions about his general health and well-being. "Our Aeromed would like to hear any comments you have on weightlessness, nausea, dizziness, taste, and smell sensations."

Glenn's response: "I have no sensations at all from weightlessness except very pleasant. No ill effects at all; I'm not sick, feel fine. I have had no, no dizziness; I've run the head checks, the head movements, and have no problem with that. I've run the ocu-logyric check and have no problem with that. I haven't been able to use much of the equipment this orbit, however. I've been mainly concerned with this control problem. Over."

Later in his communication the Canton CapCom said, "We also have no indication that your landing bag might be deployed."

When Glenn asked if someone had reported that there might be a problem with the heat shield, which enclosed the landing bag, the response did little to clear up the puzzle.

"Negative," the CapCom replied. "We had a request to monitor this and to ask you if you heard anything flapping."

The lack of information began to irritate Glenn. What did they know on the ground that they weren't telling him? Was there some indication of a serious problem with the space-craft? However he was also concerned with other issues that he could work on, and later said that with all the control system problems he had serious doubts that he would be allowed to complete a third orbit.

"The automatic controls were misbehaving; the manual controls that I had relied on most had become a little mushy – at least, they were not as crisp as they had been. With the problems we were having, I was concerned that perhaps the people down on the ground might prefer for me to come on home. I sincerely hoped not. There was nothing to be concerned about unless the problem got significantly worse. We were still in good shape and I felt that if it was going to be necessary for me to bring the capsule back myself, I might as well have another 90 minutes of practice at the controls."[18]

To his relief, as he passed over Hawaii, the ground station gave Glenn the hoped-for go-ahead to conduct a third orbit, as related in the post-flight report prepared by Flight Director Chris Kraft.

"As the go-no-go point at the end of the second [orbit] and beginning of the third orbit approached, it was determined that although some spacecraft malfunctions had occurred, the astronaut continued to be in excellent condition and had complete control of the spacecraft. He was told by the Hawaiian site that the Mercury Control Center had made the decision to continue into the third orbit. The astronaut concurred, and the decision was made to complete the three-orbit mission."[19]

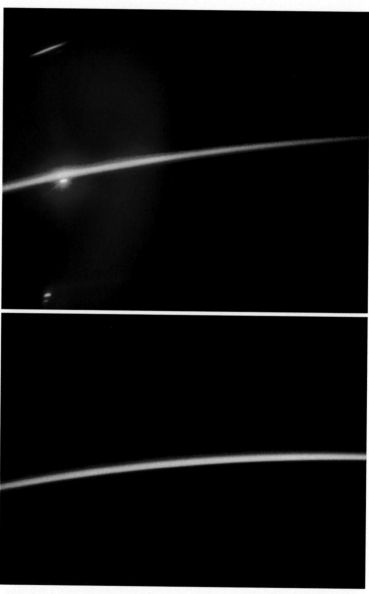

Sunset photographs taken from *Friendship 7* as it flew into a third orbit. (Photos: NASA)

"A REAL FIREBALL"

The ASCS problems continued throughout the third orbit, and when Glenn was in touch with Gordon Cooper at Muchea he reported, "ASCS is [a] major item, still not operating properly. I am on fly-by-wire at present time. I have no low thrust to the right fly-by-wire."

While he had Cooper in contact, Glenn asked if he could pass on a message for him.

"I want you to send a message to the Director, to the Commandant, U.S. Marine Corps, Washington. Tell him I have my four hours required flight time in for the month and request [a] flight chit be established for me. Over."

"Think they'll pay it?" Cooper asked, highly amused.

"I don't know. Gonna find out," was the response from orbit.

"Roger. Is this flying time or rocket time?"

"Lighter than air, Buddy," came Glenn's reply.

As the time for retrofire approached, Glenn began to sense that everything was not quite as it should be. There was further evidence of this when the CapCom in Hawaii asked him to check the landing bag deploy indicator light. Glenn obliged, flipping the switch on and off and reported that – as expected – there was no light showing. He then went back to running through his retro-sequence checklist and preparing for the retrofire maneuver.

A panoramic view of Florida taken by Glenn showing the Georgia border (right, under clouds) to just north of Cape Canaveral. (Photo: NASA)

It was agreed with ground control that they would follow the flight plan and the retro-fire maneuver would take place using the ASCS, with Glenn prepared to take over manual control should a malfunction occur. Additional time checks were also made over Hawaii to ensure that the retrofire clock was properly set and synchronized to provide retrofire at the precise moment. Glenn was asked to make a one-second adjustment to his time clock.

Although Glenn's transmission that no indicator light was showing provided a degree of reassurance to the worried controllers at the Cape, the problem existed of how they should tell the inquisitive Marine *what* to do with his spacecraft without revealing the reason *why* he had to follow their instructions. They had to move fast, and decided to frame their decision as a recommendation.

The next CapCom contact for Glenn was Wally Schirra at the Point Arguello station in California. After confirming good and clear communications, and with less than a minute to the retrograde sequence, Schirra said he would give a countdown, and added a reassuring, "You're looking good."

Glenn repeated that he was on ASCS, backing it up manually, and his fuel was 39 percent. At the 30-second mark he advised that the retro-warning light was on. After acknowledging this Schirra said, "John, leave your retropack on through your pass over Texas. Do you read?" This was acknowledged as Glenn prepared for the re-entry phase to begin. At 4 hours 33 minutes and 9 seconds into the mission flight time, he reported that the first of the retrorockets was firing right on time.

"Sure they be," Schirra acknowledged.

"Are they ever," Glenn said. "It feels like I'm going back toward Hawaii."

Schirra chuckled. "Don't do that, you want to go to the East Coast."

Five seconds later the second retrorocket kicked in, and the third another five seconds on. Each retrorocket fired for 12 seconds before burning out. The overall effect was to reduce the capsule's speed by around 330 miles an hour, which would cause it to fall below orbital speed and begin the re-entry process. Ordinarily the retro-package would be jettisoned at this point, but Schirra reminded Glenn to keep the retro-package on until he was past Texas. In other words, wait for further instructions.

After further communications about the capsule's attitude, Glenn asked Schirra if he had a revised time for retro-package jettison, but Schirra would only say, "Texas will give you that message."

Glenn had little time to concern himself with this added problem, as the automatic yaw control kept slamming the capsule back and forth. He decided to revert to manual control in all three axes. He then heard from the next communication point, Guaymas in north-western Mexico, before establishing communication with the CapCom at Corpus Christi, Texas. As soon as communications formalities had been cleared, Glenn received the information passed through to Texas from Mercury Control.

"We are recommending that you leave the retro-package on throughout the entire re-entry. This means that you will have to override the 0.05-g switch, which is expected to occur at 04 [hours] 43 [minutes] 53 [seconds]. This also means you will have to manually retract the scope. Do you read?"

"This is *Friendship Seven*," Glenn responded. "What is the reason for this? Do you have any reason? Over."

"Not at this time," came the reply. "This is the judgment of Cape Flight. . . . Cape Flight will give you the reasons for this action when you are in view."

Thirty seconds later the astronaut was in contact range of the Cape, and Alan Shepard relayed instructions for retracting the periscope. In what almost seemed like an after-thought, he quickly added, "While you're doing that, we're not sure whether or not your landing bag has deployed." Before a puzzled Glenn could respond, Shepard explained, "We feel it is possible to re-enter with the retro-package on. We see no difficulty at this time in that type of re-entry. Over."

"Roger, understand," Glenn replied, completely baffled, but knowing he could save any questions and criticism for his post-flight debriefing.

Things were about to heat up – literally. After Glenn had acknowledged Shepard's transmission about leaving the retro-package on he prepared himself for what – even under normal circumstances – was an incredibly hazardous plunge through the atmo-sphere. All he could do now was rely on his training, the integrity of *Friendship 7,* and physics, to get him back safely.

The retrorockets were positioned in a round aluminum container called a retropack (or retro-package) on the exterior of the heat shield. The pack was held firmly to the base of the capsule itself – not to the heat shield – by three stout straps made of exceedingly strong titanium.

In his post-flight report, Chris Kraft noted that even though all the test results concern-ing the Segment 51 anomaly were negative and further indicated that the signal was erro-neous, they were not conclusive. "There were still other possible malfunctions and the decision was made at the control center that the safest path to take was to leave the retro-package on following retrofire. This decision was made on the basis that the retro-package straps attached to the spacecraft and the spacecraft heat shield would maintain the heat shield in the closed position until sufficient aerodynamic force was exerted to keep the shield on the spacecraft. In addition, based on studies made in the test, it was felt that the retention of the package would not cause any serious damage to the heat shield or the spacecraft during the re-entry and would burn off during the re-entry heat pulse."[20]

Following Shepard's instruction, Glenn retracted the periscope by pumping the manual retraction lever. As the craft's deceleration rate increased he could hear a hissing noise that sounded to him like small particles brushing against the capsule.

"It was now 4 hours and 41 minutes after launch. In another few minutes, I would be in the hottest part of my ride. The automatic control system had been acting up again – drifting off center and then kicking itself back into line again – so I was now controlling almost completely by the manual stick. The fuel in the manual system was running low, however – the gauge read that I had only about 15 percent left in the tank. So I switched to fly-by-wire in order to draw on what fuel was left in the automatic system. I was still controlling manually, however, since this was the advantage that the fly-by-wire system provided. I used the manual stick; the nozzles that I activated and the fuel that I expended belonged to the automatic system."[21]

Glenn's principal task now was to keep the blunt end of his spacecraft aligned at a spe-cific attitude as it tore through the atmosphere. He knew the next few minutes were going to be the most critical and dangerous he had ever lived through. He would either come through it safely, or he would perish in the unforgiving heat which would cause *Friendship 7* to rapidly overheat and disintegrate into white-hot embers. Hence any sudden and continuing rise in temperature within the cabin would indicate that he might be in dire trouble.

Soon after that he heard a loud thump coming from behind him, and believed it to be the retro-package letting go. He quickly conveyed this information to Alan Shepard, but there was no response.

By this time, radio contact had been lost due to the effects of ionization now enveloping the capsule. Shepard had also been trying to tell Glenn to jettison the retro-package as soon as the dynamic load built up to 1 or 1.5 g's, but the radio blackout meant that the astronaut never received this information. He was now engulfed in radio silence. His reports during this time were captured by a tape recorder.

All of a sudden, Glenn noticed one of the three metal retro-package straps flapping around wildly outside his window. It was a somewhat alarming sight, but he continued to operate as calmly as possible, keeping his spacecraft under control. He then noticed "a bright orange glow" building up all around him. At 4 hours and 44 minutes into the flight he reported, "This is *Friendship 7*. A real fireball outside."

THE MOMENT OF TRUTH

Glenn was really sweating it out now. At its most ferocious, the temperature directly on the blunt end of the capsule would reach 3,000 degrees Fahrenheit, while the heat building up around the capsule could go as high as an astonishing 9,500 degrees Fahrenheit – almost as hot as the surface temperature of the Sun. He tried to remain calm, diligent, and professional in the deeply disturbing realization that if the worst possible situation occurred, there was absolutely nothing he could do about it. "I knew that if the heat shield was falling apart, I would feel the heat pulse first at my back, and I waited for it."[22]

Soon after, he saw chunks of flaming material whipping past his window along the capsule's flight path as he continued to observe the dazzling orange glow surrounding *Friendship 7*.

"It lasted for only about a minute, but those few moments ticked off inside the capsule like days on a calendar. I still waited for the heat, and I made several attempts to contact the control center and keep them informed."[23] He called repeatedly, but there was no response owing to the radio blackout caused by ionization around the capsule. Meanwhile, the forces acting on his body quickly built up to around 8 g's.

It would prove to be one of the most worrisome times in the history of American space exploration as *Friendship 7* plowed a path through the intense heat of re-entry. In the Mercury Control Center the capsule's flight path was monitored on C-band radar units while waiting for communications to be restored. The controllers, although tense, were comforted in knowing that all the data available to them indicated the capsule remained on its planned descent, and was intact. At least Alan Shepard was able to keep his anxiety somewhat in check by repeatedly trying to make contact with the spacecraft. Many silent prayers were being offered for the safety of the astronaut.

Dr. William Douglas says a quiet prayer for John Glenn as tension mounts within the Mercury Control Center. (Photo: NASA)

Finally, at 4 hours, 47 minutes and 11 seconds into the flight, communications were restored as Glenn heard Shepard's voice asking if he could read his transmission, and he smiled with relief. He knew the worst was over; he had survived re-entry, and splashdown would occur in less than eight minutes. Outside the capsule, the orange glow was fading.

"Loud and clear," he said, trying to sound calm. "How me?"

"Roger," came the response. "Reading you loud and clear. How are you doing?"

Having just passed through several minutes of the most critical time in his entire life, Glenn was bursting to tell everyone he was okay, but his reply was a surprisingly laconic, "Oh, pretty good."

Shepard quickly brought Glenn up to date with the readings, which indicated he would splash down within about a mile of one of the U.S. Navy's ships involved in the recovery operations. It was very welcome news for the astronaut.

"My condition is good," he continued to report, reassuring everyone on the ground. "But that was a real fireball, boy." It was a defining statement that would grace the front pages of many of the world's headlines soon after.

Twelve seconds later, *Friendship 7* was passing through 80,000 feet, and another 19 seconds on Glenn's altimeter read 55,000 feet.

SPLASHDOWN!

As Glenn later recorded in his pilot's report, "After peak deceleration, the amplitude of the spacecraft oscillations began to build. I kept them under control on the manual and fly-by-wire systems until I ran out of manual fuel. After that point, I was unknowingly left with only the fly-by-wire system and the oscillations increased; so I switched to auxiliary damping, which controlled the spacecraft until the automatic fuel was also expended."[24]

Glenn was reaching for the switch to deploy the drogue parachute early in order to reduce the oscillations, when it was deployed automatically.

"I could feel the thud of the mortar that launched the drogue," he recalled. "The window was covered now with a thin layer of melted resin that had streamed back from the heat shield. However, I could see the drogue open up. The swaying motion was cut sharply."[25]

As the drogue parachute helped to stabilize the spacecraft, the periscope also extended automatically. Then, as the spacecraft passed through 10,800 feet, a barometric switch set off the mortar charge that fired out the main parachute.

"Then, through the dirty window and the periscope, I saw a marvelous chain reaction set in. I watched the antenna canister go that housed the chute. It dragged the main chute along behind it, wrapped up inside its bag. When the shrouds of the chute were stretched out to their full length, the bag peeled off and left the chute, still in a reefed condition, training out like a long ribbon straight above me. Then the reefing lines broke away and the huge orange and white chute blossomed out, pulsed several times, and was steady. I could feel the hard jolt in the cabin as we slowed down."[26]

The capsule instruments indicated that the rate of descent was some 10 feet faster per second than planned, so Glenn examined the main chute carefully for any signs of tears or holes, but it seemed to be intact. He therefore dismissed any thoughts of deploying the reserve chute, which was also packed into the nose of the capsule. "It was," he said, "a moment of solid satisfaction," and he told Shepard that he had "a beautiful chute."[27]

As he continued his descent, Glenn established contact with the recovery forces and was patched through to the commander of the destroyer USS *Noa*, Capt. John Dryden Exum, who gave him an estimated landing point and said they had their direction finders trained on the spacecraft and the crew had already spotted the main chute. They were about six miles from the splashdown zone and he told Glenn the ship would take an estimated 20 minutes to reach the pickup area.

Glenn then began the process of running through his landing checklist. First, according to procedure, he unfastened a plug on the leg of his space suit which was connected to his body sensors. He then removed the blood pressure equipment from the suit, loosened and freed his chest strap, disconnected the visor seal hose, unlocked the helmet visor and pushed it back. Next, he unhooked the respiration sensor from his lip mike and stuffed it inside his suit, and disconnected the helmet hose. Lastly he retrieved the capsule's emergency survival pack from the left side of his couch, in case it was needed.

Shepard then asked Glenn if he had a green light on the landing bag deployment. This bag, which now extended several feet below the capsule, was designed to soften the shock of landing on the water. Glenn confirmed that the landing bag light was green, which meant that it had successfully deployed. Shepard then passed to the *Noa* and Glenn the information that he should remain in his spacecraft for the pickup, unless he had an "overriding reason" to get out. Glenn acknowledged this and then prepared himself for the moment of impact with the water. This would occur at 2:43 p.m. (EST) after 4 hours, 55 minutes and 23 seconds of flight.

"The capsule hit the water with a good solid bump and went far enough under water to submerge both the periscope and the window. I could hear gurgling sounds almost immediately. After it listed over to the right and then to the left, the capsule righted itself and I could find no traces of any leaks. I undid the seat strap now and the shoulder harness, disconnected my helmet and put up my neck dam so I could not get water inside my suit if I had to get into the ocean. I was sweating profusely and was very uncomfortable."[28]

For a few moments, contrary to instructions, Glenn considered removing the lid on the neck of the capsule and climbing out, as he had practiced in training sessions. Despite his discomfort he decided against this, unwilling to generate even more heat by such exertions.

"The thing to do now was sit tight."[29]

#

In regard to the heat shield Segment 51 anomaly, technicians later discovered the cause of the erroneous signal. It was nothing more than an improperly rigged switch. The heat shield and the astronaut had never been in jeopardy.

REFERENCES

1. Guenter Wendt and Russell Still, *The Unbroken Chain*, Apogee Books, Ontario, Canada, 2001
2. Dee O'Hara email correspondence with Colin Burgess, 23 November 2014
3. Guenter Wendt and Russell Still, *The Unbroken Chain*, Apogee Books, Ontario, Canada, 2001
4. Donald K. Slayton with Michael Cassutt, *Deke! U.S. Manned Space: From Mercury to the Shuttle*, Forge Books, New York, NY, 1994
5. John Glenn, article, "If you're shook up, you shouldn't be there," from *Life* magazine, issue 9 March 1962, pp. 25–31

6. John Glenn with Nick Taylor, *John Glenn: A Memoir*, Bantam Books, New York, NY, 1999
7. John Glenn, article, "If you're shook up, you shouldn't be there," from *Life* magazine, issue 9 March 1962, pp. 25–31
8. *Ibid*
9. *Ibid*
10. Carpenter, S., Cooper, Jr. L, Glenn, Jr., J., Grissom, V., Schirra, Jr., W., Shepard, Jr., A., and Slayton, D., *We Seven*, Simon and Schuster Inc., New York, NY, 1962
11. John Glenn with Nick Taylor, *John Glenn: A Memoir*, Bantam Books, New York, NY, 1999
12. Carpenter, S., Cooper, Jr. L, Glenn, Jr., J., Grissom, V., Schirra, Jr., W., Shepard, Jr., A., and Slayton, D., *We Seven*, Simon and Schuster Inc., New York, NY, 1962
13. John Glenn, article, "If you're shook up, you shouldn't be there," from *Life* magazine, issue 9 March 1962, pp. 25–31
14. Gene Kranz, *Failure is Not an Option*, Simon and Schuster Inc., New York, NY, 2000
15. Manfred (Dutch) von Ehrenfried email correspondence with Colin Burgess, 28–30 October 2014
16. *Ibid*
17. Gene Kranz, *Failure is Not an Option*, Simon and Schuster Inc., New York, NY, 2000
18. Carpenter, S., Cooper, Jr. L, Glenn, Jr., J., Grissom, V., Schirra, Jr., W., Shepard, Jr., A., and Slayton, D., *We Seven*, Simon and Schuster Inc., New York, NY, 1962
19. *Ibid*
20. John Glenn, "Pilot's Flight Report," taken from *Results of the First United States Manned Orbital Space Flight, February 20, 1962*, NASA Manned Spacecraft Center, Houston, TX, 1962
21. Christopher C. Kraft, Jr., "Flight Control and Flight Plan," taken from *Results of the First United States Manned Orbital Space Flight, February 20, 1962*, NASA Manned Spacecraft Center, Houston, TX, 1962
22. Carpenter, S., Cooper, Jr. L, Glenn, Jr., J., Grissom, V., Schirra, Jr., W., Shepard, Jr., A., and Slayton, D., *We Seven*, Simon and Schuster Inc., New York, NY, 1962
23. *Ibid*
24. John Glenn, "Pilot's Flight Report," taken from *Results of the First United States Manned Orbital Space Flight, February 20, 1962*, NASA Manned Spacecraft Center, Houston, TX, 1962
25. John Glenn, article, "If you're shook up, you shouldn't be there," from *Life* magazine, issue 9 March 1962, pp. 25–31
26. *Ibid*
27. Carpenter, S., Cooper, Jr. L, Glenn, Jr., J., Grissom, V., Schirra, Jr., W., Shepard, Jr., A., and Slayton, D., *We Seven*, Simon and Schuster Inc., New York, NY, 1962
28. John Glenn, "Pilot's Flight Report," taken from *Results of the First United States Manned Orbital Space Flight, February 20, 1962*, NASA Manned Spacecraft Center, Houston, TX, 1962
29. *Ibid*

N.B. *Mission voice communications taken from NASA transcript of MA-6 air-to-ground transmissions*

7

Safe splashdown

Operating under the code designation "Steelhead," a 390-foot Atlantic Fleet *Gearing*-class destroyer and her crew were destined that day to become famous around the world. The USS *Noa* (DD-841), named for Midshipman Loveman Noa, was normally attached to DESDIV (Destroyer Division) 142 with her home port in Mayport, Florida. However, on 20 February 1962, the 17-year-old destroyer was operating as a member of a special recovery task force, ready to pluck the *Friendship 7* spacecraft and astronaut John Glenn out of the Atlantic after their epic three orbits around the Earth. Despite being one of many ships standing by for the splashdown, *Noa*'s crew and her Tennessee-born skipper, Commander John Exum, never lost faith that their ship had as good a chance as any to carry out the recovery. They made plenty of advance preparations on the strength of that faith. And then, they got lucky.

LUCK OF THE DRAW

As part of the Mercury recovery fleet, the USS *Noa* was one of three ships assigned to patrol an oval-shaped area some 200 miles long and 50 miles wide. The center of this particular area, some 950 miles southeast of Cape Canaveral and roughly 225 miles northwest of San Juan, Puerto Rico, was designated as the prime recovery area in the event that three Earth orbits were accomplished. Right in the middle of this zone was the anti-submarine warfare aircraft carrier, USS *Randolph* (CVS-15). *Noa*, meanwhile, had been ordered to maintain a position approximately 50 miles astern of *Randolph*, while the destroyer USS *Stribling* (DD-867) steamed a similar distance off the carrier's bow. Other ships waited and watched, too. If only two orbits were achieved, the task of recovery would have fallen to another three ship force – the aircraft carrier USS *Antietam* (CVS-36) and destroyers USS *Bailey* (DD-492) and USS *Turner* (DDR-834), which were positioned in an identically shaped area approximately 360 miles southeast of Bermuda.

© Springer International Publishing Switzerland 2015
C. Burgess, *Friendship 7*, Springer Praxis Books, DOI 10.1007/978-3-319-15654-5_7

USS *Noa* at sea. (Photo: U.S. Navy)

In a single-orbit situation, *Friendship 7* would have landed somewhere in a like area 600 miles east of Bermuda, where the attack aircraft carrier USS *Constellation* (CVA-64), USS *Norfolk* (DL-1) and USS *Stormes* (DD-780) were deployed. Beyond and surrounding these three recovery areas, the rest of the 24-ship task force, including 60 aircraft, plus many shore-based patrol, search, and rescue planes were also on the alert.

Under the primary recovery plan, with *Friendship 7* expected to splash down some-where close to the *Randolph*, a Marine Corps HUP-2 helicopter from that ship would recover Glenn, expected to be sitting atop the floating spacecraft, and then either the *Noa* or *Stribling* would hoist the unoccupied capsule aboard. But recovering the astronaut him-self was the primary ambition of every ship's crew that day, and there was a lot of rivalry and hopeful expectation. As the *Noa*'s Executive Officer, Lt. William Hatcher, stated prior to the space shot:

"We feel that the capsule is far more likely to undershoot than overshoot. We are bank-ing on it falling a little short. If it falls as much as 40 to 50 miles short, we stand a good chance of beating those Marine choppers to the punch. The colonel won't want to stay in that capsule any longer than necessary once it hits the water. He'll want to get off on the first thing that comes by."[1]

Like all of the destroyers in the recovery task force, the *Noa* was well prepared for the possibility of recovering the spacecraft from the water. One of her whaleboat davits had been specially rigged with a large winch and a super-strong braided nylon line. And, just in case, a large blue-and-white sign had been painted and mounted on the ship's deck, proclaiming "*Noa*'s men welcome Glenn." The ship's captain had also made some arrange-ments should they happen to be the ones to welcome Glenn aboard; he had organized to have a citation prepared, naming Glenn as an honorary member of the ship's company, and *Noa*'s sailor of the month. As the latter award normally came with a check for $15, this had already been signed and was in the ship's safe, ready to be presented.

Sailors from the USS *Noa* with their hopeful sign. (Photo: U.S. Navy)

ACTION ON THE USS *NOA*

Coming at 2:40 p.m., 4 hours and 53 minutes after lift-off, *Friendship 7* announced her safe arrival through the Earth's atmosphere with a huge sonic boom, which was clearly audible aboard the *Noa*. Soon after, Seaman Vaughn Helms, one of a veritable swarm of sailors topside on the ship's deck and superstructure, was the first to spot the spacecraft, dangling below a huge orange parachute a little over three miles away as it descended for a watery climax to its historic journey into space.

Another who spotted the spacecraft from the *Noa* was Donald Harter from Columbus, Ohio, from the USS *Randolph* Command Anti-Submarine Helicopter Squadron HS-7. He had been transferred across to the *Noa* two weeks earlier to attend recovery briefings. "I was a lookout using big powerful long-range observation binoculars, knowing Glenn was on his third and last orbit, looking for his sixty-three-foot red and white parachute which was described in the briefings. I heard the sonic boom, sighting the main parachute at an altitude of about 5,000 feet, from a range of three nautical miles."

Even before the spacecraft splashed down at 2:43 p.m., amid a sudden, hissing cloud of steam, three Marine Corps HUP-2 helicopters from the *Randolph* were airborne and rushing to the scene. So too, was the *Noa*, and the gallant destroyer would win the race.

"Glenn was able to talk to us, his recovery team by radio," Harter recalled. "When we arrived a bright light was flashing on top of the spacecraft, a smoke bomb had ejected that poured brightly colored smoke downwind. Also a dye was released which stained the water, and a sonar depth charge exploded to set off shockwaves that could be detected by listeners expecting it aboard the ships nearby. He radioed that he was waiting to be picked up."[2]

The aircraft carrier USS *Randolph*, February 1962. (Photo: U.S. Navy)

Friendship 7 bobbing in the water as the USS *Noa* draws near. (Photo: NASA)

As they neared the spacecraft, with communication established with the astronaut, Capt. Exum guided the *Noa* into an approach from the windward side, which enabled the ship to move alongside *Friendship 7*. With the nearest helicopter still some 15 miles distant, the *Randolph* – somewhat grudgingly – gave permission for the *Noa* to carry out the recovery of the astronaut and his spacecraft.

Helicopter crews aboard the USS *Randolph* prepare to fly out to the splashdown site. (Photo: NASA)

As *Friendship 7* rolled back and forth in the ocean swells, Glenn said he had received a running commentary from the *Noa* as the ship rapidly closed in. "First, she was 4 minutes away, then she slowed down and was 3 minutes away; then her engines were stopped and she was coming alongside. The capsule window was so clogged now with both resin and sea water that I could not see her. Strangely enough, however, the capsule bobbed around in the water until the periscope was pointing directly at the destroyer, and it kept her in view from then on. I could read her number – 841 – and I could see so many sailors in white uniforms standing on the deck that I asked the captain if he had anybody down below running the ship. He assured me he did."[3]

As recalled by U.S. Naval Rating Jerry McConnell, JO1 (Journalist First Class), a special sea detail on the *Noa* under the direction of Lt. (junior grade) James Herr then swung into action.

"Wielding a shepherd's crook, Boatswain's Mate 3rd Class David Bell and Seaman Earnest Knowles leaned out and snagged the capsule. Then the super-strength nylon line was run through a metal eyelet on the capsule's top and, with the winch hauling away, and with such *Noa*men as Ensign Donald Batista, Chief Boatswain's Mate Harold L. Isehower and Joe Nelson, BM2, playing key roles, the *Friendship Seven* was hoisted easily onto *Noa*'s main deck." *Friendship 7* was placed on the deck at 3:04 p.m., some 21 minutes after splashdown.

"During the three-mile run to the scene, and throughout the entire pickup operation, *Noa* was in constant radio contact with astronaut Glenn inside the capsule, and had repeatedly warned him not to attempt to come out. Now however, safely ensconced on board, Colonel Glenn wasted little time blowing the escape hatch in the capsule's side and clambering out to a tumultuous welcome from *Noa*'s crew."[4]

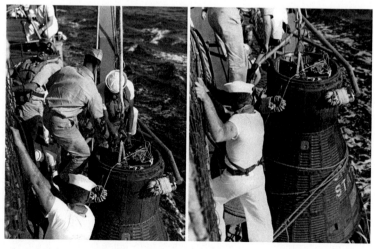

Sailors aboard the *Noa* assist as *Friendship 7* is hoisted aboard the destroyer. (Photos: NASA)

Friendship 7 is successfully brought aboard the USS *Noa*. (Photo: U.S. Navy)

ASTRONAUT ON BOARD

At first, Glenn said he had started to crawl through the top end of *Friendship 7* to avoid blowing the side hatch "and jiggling the instruments inside the capsule. I was still so uncomfortably hot, however, that I decided there was no point in going out the hard way. After warning the deck crew to stand clear, and receiving clearance that all of the men were out of the way, I hit the handle which blew the hatch. I got my only wound of the day doing it – two skinned knuckles on my right hand where the plunger snapped back into place after I reached back to hit it. Then I climbed out on deck. I was back with people again."[5]

According to Jerry McConnell: "His first words – 'It was hot in there' – were hardly historic, but certainly understandable."[6]

One of the first people to greet Glenn was Boatswain's Mate David Bell. "Nice going, Colonel Glenn. Glad to have you back," Bell recalled saying. "I'm from Ohio, too." Glenn replied with a smile, "Thanks a million. Never was so glad to see anything in my life as this ship."[7]

The moment news was relayed that Glenn was safely on board the *Noa*, the Post Office Department in Washington, D.C., issued orders to postmasters across the United States to unseal the tightly wrapped packages that had been delivered to them weeks before. Inside were thousands of four-cent Project Mercury stamps that had been secretly printed for just this moment, and they went on immediate sale.

After blowing the hatch on his capsule, Glenn was assisted out onto the deck of the *Noa*.
(Photos: Associated Press)

In a later, post-flight photo, John Glenn holds a sheet of the specially printed stamps. (Photo: NASA)

Electrician's mate 2nd Class Bob Frangenberg recalls seeing Glenn finally emerge from the spacecraft. "He crawled out through the hatch, which was the window on the capsule, and he was soaking wet; absolutely drenched in sweat."

Despite his euphoria at being on board, Glenn was uncomfortably hot and clammy inside his space suit and wanted to doff it at the first opportunity. He was escorted to the captain's cabin where the two flight surgeons helped him to slip out of the sleeves. "That felt pretty good. I peeled off the rest of the suit and felt still better." He then removed the biosensors and the urine collection device. It was only when he had taken off his spacesuit that Glenn realized the hatch's firing ring had kicked back and skinned his knuckles. "Then, in a pair of long-handled space skivvies which were wringing wet with perspiration, I stepped out onto the deck and stood in the breeze. That felt best of all."[8]

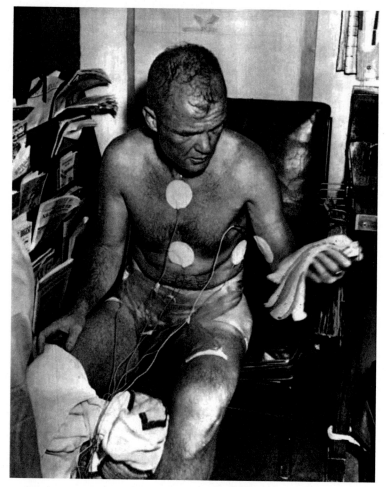

Having stripped off his spacesuit and undergarment, Glenn removes the sensor devices from his body. (Photo: Associated Press)

THE PRESIDENT CALLS

Lt. Cdr. Robert Mulin and Army physician Capt. Gene McIver were the doctors on board *Noa*, but they were not equipped to take x-rays or an electrocardiogram, and the ship was rolling too much to assess Glenn's weight accurately. After the astronaut had showered, and knowing they needed to record some immediate medical data they conducted a pre-liminary examination after agreeing that a more comprehensive one would be conducted once he was aboard the *Randolph*, which was currently steaming towards the destroyer. Later on, they would describe Glenn as being "hot, sweating profusely, fatigued, lucid but not talkative. Following a glass of water and a shower, he became more loquacious. He admitted only to some 'stomach awareness,' beginning after he was down on the water but had experienced no nausea or stomach unease during the flight. Due to dehydration, he had lost 5 pounds, 5 ounces from his pre-flight weight of 171 pounds, 7 ounces. An hour after landing his temperature was 99.2 degrees, only a degree higher than his pre-flight reading."[9]

After drinking a large glass of iced tea, the cooled-down astronaut was relaxing in the captain's cabin when he received a brief but heartfelt radio telephone call from President Kennedy.

The President:	"Hello?"
Glenn:	"Hello, Sir."
The President:	"Colonel?"
Glenn:	"This is Colonel Glenn."
The President:	"Listen, Colonel, we are really proud of you and I must say you did a wonderful job."
Glenn:	"Thanks, Mr. President."
The President:	"We are glad you got down in very good shape. I have just been watching your father and mother on television and they seemed very happy."
Glenn:	"It was a wonderful trip – almost unbelievable thinking back on it right now. But it was really tremendous."
The President:	"Well I am coming down to Canaveral on Friday and hope you will come up to Washington on Monday or Tuesday and we will be looking forward to seeing you there."
Glenn:	"Fine. I will certainly look forward to it."

Shortly after that, Glenn also called his excited wife, Annie. She was both thrilled and relieved. Mostly relieved (although she didn't let on to her husband at the time), as Scott Carpenter had phoned during the final phases of the flight in regard to the possibility of the heat shield being loose, and to prepare Annie just in case her husband didn't make it back. With these calls and formalities over, and while waiting to be picked up by helicopter for transfer across to the *Randolph*, Glenn opened a kit that had been sent ahead to each of the recovery ships with personal equipment that he might need until he could make it to Grand Turk Island where his own gear was waiting for him.

"The kit included underwear, dark glasses, a wrist watch, a flying suit, a pair of sneakers and a blank check. I assumed that the check was to be made out to my own bank account to provide me with some spending money until I could get home. I had already thought of that, however, and had given Scott Carpenter some money to hold for me until we caught up with each other at Grand Turk. I did not fill out the check."[10]

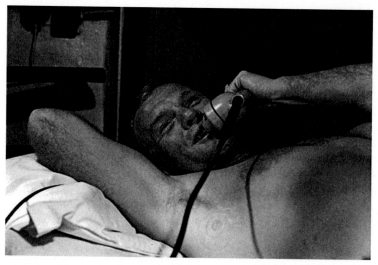

Glenn receives his congratulatory phone call from President Kennedy. (Photo: NASA)

President Kennedy calling America's newest hero from the White House. (Photo: John F. Kennedy Museum Photo Library)

Now comfortably dressed in a blue flight suit, Glenn prepares to record some initial impressions of his flight. (Photo: NASA)

A SHIP CELEBRATES

Now changed into a light blue flight suit and high-top sneakers, Glenn made his way out onto the deck and sat down on a chair on *Noa*'s fantail and propped his feet up on a handy rig on what was called the hedgehog deck. Then, speaking into a tape recorder, he carefully conducted his own self-briefing session while the details and impressions of his flight were still fresh in his mind. When he was finished, the ship's commanding officer John Exum called all hands together for a brief ceremony.

John Glenn packs up his spacesuit for shipment back to the Cape. (Photo: NASA)

Before this began, Glenn asked to make an announcement over the ship's loudspeaker in which he thanked the crew. He promptly won their hearts by declaring "there's not another ship in the U.S. Navy I'd rather be a crew member of." Then there was the reading of the citation naming him as an honorary crew member of the *Noa*, along with the presentation of the $15 check for his unanimous selection as *Noa*'s Sailor of the Month for February, which was a somewhat unprecedented award for a U.S. Marine. His popularity soared even more when he endorsed the check over to the ship's Welfare and Recreation Fund (who promptly said it would be framed).

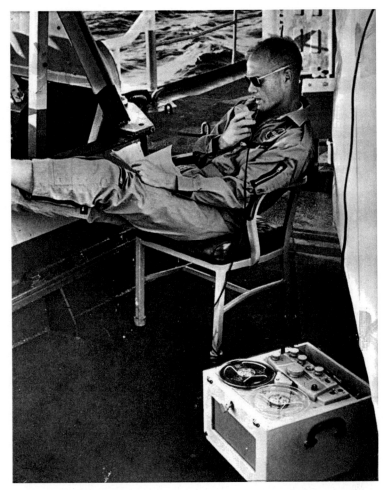

Glenn recounts his flight into a tape recorder in a quiet spot on the ship. (Photo: NASA)

"The chaplain on the destroyer presented me with a Bible, which he said was a 'navigation chart to the heavens which I had just visited.' It was a touching ceremony … I found out later that to commemorate the occasion the crew painted some white footprints where my boots hit when I jumped out."[11]

According to Navy journalist Jerry McConnell, it was almost time for the astronaut to leave the ship.

"All of this time *Randolph*, carrying some of the most frustrated newsreel and television cameramen and reporters in the history of newsgathering, was speeding towards *Noa* at a flank-speed 30-plus knots. For a while there was talk of a highline transfer of the astronaut between *Noa* and the *Randolph* when the two ships rendezvoused, but this idea was quickly scuttled in favor of a less risky and less elaborate helicopter pick-up.

Curious crewmen peer at Glenn from an upper deck. (Photo: NASA/Dean Conger)

"At about 1730, a *Randolph* chopper hovered over *Noa*'s deck, a sling was lowered, and Colonel Glenn was lifted up and in. The almost three-hour tour of sea duty aboard *Noa* was over for astronaut Glenn. Ahead, for him, lay a couple of more hours of medical exams and further debriefing sessions aboard *Randolph*, more of the same at Grand Turk Island in the Bahamas, then a return to the U.S. for one of the wildest welcomes ever accorded any American.

"Ahead for *Noa* lay a return to Mayport, and a resumption of her more prosaic every-day antisubmarine warfare training duties. For a few short hours, though, she and her crew had basked in the glory of a central role in one of America's first great thrusts into space."[12]

As Glenn remembered about leaving the *Noa*, "Just before sundown a helicopter arrived from the carrier *Randolph* to transfer me to that ship. I wanted to see the capsule once more before I left so I went back on the deck where it was tied in place and retrieved the ditty bag that was still lying inside it, which contained instruments, camera and film. The Sun was setting now, and I took a good look. It was the fourth beautiful sunset of the day, and you don't often have days like that."[13]

MEMORIES OF AN HISTORIC DAY

Eugene ("Gene") Wolfe is a retired U.S. Navy signalman first class, and he has many fond recollections of the day John Glenn was welcomed aboard the USS *Noa*.

"I was a signalman on board the *Noa* and I was on the signal bridge when we picked up the capsule containing Glenn using the starboard boat davit. The carrier USS *Randolph* was originally assigned to pick him up by helicopter out of the ocean, but when he under-shot the landing site a little we were first to reach him.

"Mr. Glenn didn't stay aboard very long; just long enough to have the medical staff look him over (all the checking was done in the wardroom I believe). And then the carri-er's helicopter came to pick him up. After he was safely aboard the helo on his way over to the *Randolph* we left the station we were patrolling and went back to Mayport, Florida, our home port. I believe the next day or two the skipper had the boatswain's mate paint on the deck where the capsule had sat after bringing it aboard. I know the marking was still on there when I left the ship for another duty station.

"I also remember that each member of the ship's crew was given an envelope with a picture of the *Noa* with the capsule on it. As far as I know we were the only tin can [destroyer] to ever pick up one of the Mercury astronauts."[14]

Sailors who have gone directly to a base, station, or ship without any specialized train-ing are eligible to select a career field. Then, through correspondence courses and exten-sive on-the-job training, they may qualify for a rating. This process is known in the U.S. Navy as "striking for a rate." If an enlisted sailor has qualified for a rate, but has not yet become a petty officer, he is called a designated striker. One such seaman, Richard Pomfrey, was aboard the USS *Noa* on 20 February as a striker, and he related to the author his excitement at having an involvement in the events of that day.

"How can I ever forget the recovery of Glenn and *Friendship 7*? Never. I was a Hospital Corpsman striker, because I hadn't been to school yet, and was straight out of boot camp on temporary sea duty. So my designation was HA, or hospitalman apprentice.

"I worked in sickbay on the *Noa* with my friend Duane Kinninger. We were both involved with securing Glenn's space suit and measuring his urine, among other minor duties. Duane and I were in the captain's wardroom performing these tasks while they were debriefing Glenn elsewhere on the ship. Being alone with Glenn's space suit, we took pictures of ourselves with his space helmet on. I can also remember transferring Glenn's urine from the catheter bag to smaller bags that were put on ice to be sent to Grand Turk Island with the other equipment.

"As Glenn was leaving the ship, I photographed him as he waved and was reeled up into a helicopter and whisked off to the carrier, *Randolph*. I'll never forget seeing the helicopters flying around overhead taking pictures of us for *Life* magazine and the newspapers. We sent letters home with the *Friendship 7* capsule stamp canceled on the *Noa* on February 20th, 1962."[15]

MORE MEMORIES

Following the launch of *Friendship 7*, a number of journalists chosen for the primary press pool had boarded a C-130 transport plane at the Cape for the two-hour flight to Grand Turk Island, where Glenn would be taken for several days of medical checkups and debriefing. Meanwhile, the *National Geographic*'s Dean Conger, on loan to NASA, was bound for the USS *Noa*. The celebrated photographer had previously captured stunning images of the post-flight recovery of Alan Shepard and Gus Grissom, and he was determined to keep the ball rolling.

As he told the author, "A NASA cinematographer and I were flown over to the *Noa* by helicopter and lowered to the bobbing destroyer's deck. A caption that later appeared in the *National Geographic* saying I arrived on the *Noa* before Glenn is not quite accurate; he was already on board. Altogether, I was on the *Noa* for maybe two or three hours. It was a hectic day." Conger was escorted to the sick bay, where Glenn was just beginning his preliminary health check. Anybody else would have been denied access, but Glenn recognized Conger from many previous photo shoots for NASA and the *National Geographic*, and was quite happy to allow him to take photographs during the examination. "After he was cleaned up a bit, John Glenn relaxed on deck and talked into a tape recorder, setting down on tape his impressions of the flight. At sunset he went to the capsule to get his ditty bag." The photo that Conger took of that event would later appear on the front cover of *Life* magazine.

"It was getting dark when he was lifted off the *Noa* by helicopter. Then I was also lifted off in another chopper and flown back to the *Randolph*, ready to record his arrival and subsequent movements on the ship. Several of us were then flown to the Cape where the world's press was anxiously awaiting photos." Dean Conger's extraordinary images taken that day would soon be seen in wire service photographs in newspapers and magazines all around the world.[16]

Tom ("Hank") Hanko was a Yeoman Third Class that day, serving on the *Noa* from 1961 to 1963. "When I boarded the *Noa* for the first time I had no idea the mission and the ship I was about to embark on would soon become one that would go down in the history books of this country. She had become a part of a mission that would forever change how the world viewed human space flight. Unfortunately I did not have any interaction with John Glenn as I was on a different part of the ship when all this was going on.

"I might add that later in life, beginning in 1984, I began working on the space shuttle program at Vandenberg AFB in California and later transferred to the Kennedy Space Center in Florida. My job at KSC, until my retirement in 2005, was a Logistics Manager in Shuttle Processing Support. I had the opportunity to meet up with Mr. Glenn again when he was at KSC preparing to make history for the second time. In October 1998 he

A *National Geographic* photographer captured this sunset picture of Glenn retrieving his ditty bag from *Friendship 7*. The photo would later feature on the front page of *Life* magazine. (Photo: Dean Conger/National Geographic/NASA)

became the oldest person to fly on the space shuttle when at age 77 he flew on the *Discovery* on mission STS-95."[17]

Bob Frangenberg, then a 19-year-old electrician's mate, had further memories to share with the author. "Back then I had a little 35-mm camera, but unlike now there was nothing automatic on it, which meant that if you wanted to focus and make a shot clearer you had to manually move some apparatus on the camera. I'd gone up to the deck where the radio shack was located, and I was standing outside when Colonel Glenn came out. So there I am walking backwards, trying to focus, and he was walking towards me. Finally he stopped and said, 'Okay, just go ahead and take your picture.' And after I'd taken the snap he moved on. I later had the pictures processed and got them back as 35-mm slides.

Dean Conger also took this iconic photograph of John Glenn at sunset aboard the USS *Noa*. (Photo: Dean Conger/National Geographic/NASA)

The one I'd taken of Colonel Glenn was really cool. I had them in a little box which I had in my coat pocket one day when I went to the ship's barber for a haircut. Afterwards I remembered I'd left my coat in the barber shop and went back. The coat was still there, but the box of slides was missing, and I never did find out what had happened to it.

"There is another story. I've read where Colonel Glenn, just before he left the *Noa*, said something about it being the fourth sunset he'd seen that day. That's quite true – in a sense. He was standing on the fantail with a few of the crew hanging around and as he looked at the sunset – which was a darned pretty sunset – he said with a laugh, and I quote, 'That's the worst f…ing sunset I've seen all day.' We loved it!

"One last story to tell about the capsule after John Glenn had left us for the *Randolph*. I can't remember the name of the person involved, but he was an IC [interior communications electrician] man. In the middle of the night he went up to the capsule, which was tied down to the deck but completely unguarded. He quickly broke off one of the tubular antennas and took it down to the machine/electrician shop where he cut it into a number of rings, which he later sold as souvenirs of the flight at $20 a pop. I don't have one, but I guess out in the world right now are several ex-sailors wearing rings made from a *Friendship 7* antenna."[18]

ABOARD THE USS *RANDOLPH*

Airman Robert ("Bob") Bell was on board the *Randolph* that day, as he had been for the Gus Grissom recovery the previous year, and he compared Grissom's rather hurried excursion on the carrier with that of John Glenn.

Crewmembers on the USS *Randolph* line the deck, ready to greet the astronaut. (Photo: U.S. Navy)

With a happy wave, Glenn walks across the hangar deck of the *Randolph*, still clutching his ditty bag. (Photo: Associated Press)

One official duty Glenn had to perform aboard the USS *Randolph* was to sign a document verifying his space flight for the records of the *Fédération Aéronautique Internationale*. Their representative (in white shirt beside him), was A. E. Hansen from the National Aeronautic Association. (Photo: NASA)

"When John Glenn's helo landed on deck, it seemed like most of the ship's crew and air group was massed to greet him. He stepped from the helo with a wave and a BIG smile. If I remember right, he walked along the line of men slowly as he made his way to the island hatch, shaking hands and speaking to them as he went. He had the ship's press make up small cards with the date and a brief note from him which he signed and gave out to the men on board the ship. It was clear he was glad to be back on Earth. And it was also clear he was not a stuffy officer type who didn't want to mingle with us lowly enlisted."[19]

Once aboard the *Randolph*, the ship's physicians took an EKG and chest x-ray of the astronaut. He would not remain on the carrier for long before a Navy S2F patrol plane was fuelled and ready to take him to Grand Turk Island, a 12-square-mile island in the Bahamas, where he would undergo a far more extensive medical examination and two days of intense debriefing, physical and psychological tests, and an engineering review of the flight.

ACTION ON GRAND TURK ISLAND

On arrival at the island around 9:30 p.m., Glenn was met by Deke Slayton and backup pilot Scott Carpenter, who greeted him with laughter, handshakes, and backslapping.

As he was escorted to a car that would convey him to the island's small hospital, the handful of designated news-pool representatives shouted out questions, asking how he felt. "Fine, wonderful!" he responded. "I couldn't feel better." Asked how he would sum up his day he said, "Well, it's been a long day – and an interesting one too." With that,

Glenn receives the plaque of Task Force Alpha from its commander, Rear Admiral Earl R. Eastwold, aboard the USS *Randolph*. Note the small bandage on Glenn's middle finger, left hand, sustained from the recoil when he punched the hatch-jettison button inside *Friendship 7*. (Photo: NASA)

he was whisked off for his medical checkup and debriefing at the hospital. After enjoying a steak dinner, his physical examination was completed by Dr. Bill Douglas, and once that was over the weary astronaut was finally able to retire to bed. Douglas stepped outside where a few reporters were gathered, and announced that, "John is in excellent condition."

On arrival at Grand Turk Island, Glenn is greeted by fellow Mercury astronauts Deke Slayton and Scott Carpenter. Col. Powers takes notes in the foreground. (Photo: NASA)

Over the next two days, Glenn underwent a battery of physical, psychological and medical tests, with plenty of rest periods thrown in. There were also a number of debriefing sessions, as he later described in his 1999 memoir.

"At the debriefing sessions, I had the highest praise for the whole operation, the training, the way the team had come back from all the cancellations, and the mission itself – with one exception. They hadn't told me directly their fears about the heat shield, and I was really unhappy about that. A lot of people, doctors in particular, had the idea that you'd panic in such a situation. The truth was, they had no idea what would happen. None of us were panic-prone on the ground, or in an airplane or in any of the things they put us through in training, including underwater egress from the capsule. But they thought we might panic once we were up in space and assumed it was better if we didn't know the worst possibilities.

"I thought the astronaut ought to have all the information the people on the ground had, as soon as they had it, so he could deal with a problem if communications was lost. I was adamant about it. I said, 'Don't ever leave a guy up there again without giving him all the information you have available. Otherwise, what's the point of having a manned program?'"[20]

Glenn later described his two days on Grand Turk Island as "an interlude, in which morning medical checks and debriefing session were followed by afternoons of play."[21]

On the second day, 22 February, as Glenn enjoyed some recreational spearfishing off a couple of glass-bottom boats along with Scott Carpenter, Bill Douglas and Wes Vickery, a technician attached to an auxiliary U.S. Air Force base on the island, the two astronauts became involved in a life-and-death incident. It began when measured descending lines were dropped over the side of the boats to gauge the depth of the water. A keen diver,

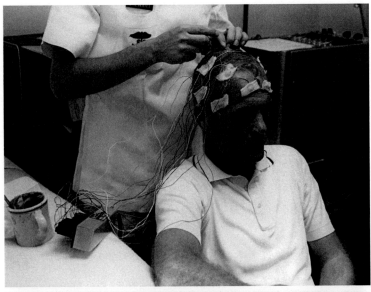

Glenn undergoes a full range of tests on Grand Turk Island. (Photos: NASA)

The morning after his arrival on the island, Glenn lines up for a hearty breakfast. (Photo: Associated Press)

Vickery asked Scott Carpenter for his help in setting a personal dive record just holding his breath. Carpenter agreed, describing the attempt by saying, "He wanted to see if he could get to a hundred feet and go back up. I was there as his safety diver."

Carpenter then donned his air tanks and mask and slipped over the side of the boat, and slowly made his way down to the 100-foot mark. When all was ready, Vickery entered the water, took a few deep breaths, then duck-dived down, following the descending line as Carpenter watched from below. "I saw Vickery coming down the line. Then, at about the fifty to sixty-foot mark, he just stopped moving. So I swam up." Vickery was unconscious, so Carpenter jammed his mouthpiece into the unresponsive diver's mouth and hauled him to the surface. Bill Douglas and a boat crewman dragged the unconscious man into their boat, where Douglas began resuscitation efforts. Finally, Vickery responded. "He came around, and he was okay," remembered Carpenter. In relating this

Scott Carpenter assists Glenn as they dive near Grand Turk Island. (Photo: NASA)

story to the author, he spoke with a quiet satisfaction, saying, "Only time I know of that I saved somebody's life!"[22]

The press soon got wind of the story, but with John Glenn being the man of the moment, newspapers inevitably carried the headline, "John Glenn Saves Diver's Life!" While Glenn was certainly in the vicinity, the real hero on that occasion was Scott Carpenter. But given the newsworthy status of the much-admired Marine astronaut, it was hardly surprising that Glenn received the credit for the rescue.

At a celebratory party held that evening at the beachside Conch Club, everyone let their hair down a little in honor of John Glenn and his successful Mercury mission. By this time Alan Shepard, Gus Grissom and Wally Schirra had all made it to the island (Gordon Cooper was still busy packing to come home from Muchea in Western Australia). Everyone wanted to meet Glenn, shake his hand, and get his autograph, and he happily obliged. During the night Bill Douglas gave Glenn and Carpenter the welcome news that Wes Vickery was fine and would make a full recovery.

A happy reunion of the Mercury astronauts on the island. Gordon Cooper was still making his way back from Australia. (Photo: NASA)

Two days later, Glenn would set the record straight when he publicly acknowledged that it was his backup, Scott Carpenter, who actually saved the diver's life. "That was impressive," Glenn said. "When you have an aqualung and you're down at about eighty-foot depth and you're trying to give your own air to someone else to help them out, that is almost heroic in my book."[23]

Meanwhile, with all the attention on the astronaut, the spacecraft that had carried him around the world three times was making a far less publicized journey of its own. The USS *Randolph* transported *Friendship 7* to Grand Turk Island. Once there it was transferred to a launch which carried it across to Grand Turk dock. The spacecraft was then craned onto a low loader, secured, and driven to the island's airport where a specially designed protective cradle was waiting. The capsule was carefully lowered into the cradle, secured, and loaded aboard a U.S. Air Force plane for its return journey to Florida.

RECALLING THREE DAYS OF EXCITEMENT

On 20 February, as the world watched their television sets or listened to radio reports of the flight of *Friendship 7*, Joe Frasketi was at a Project Mercury tracking station on Grand Turk Island, working at the telemetry site operating electronics equipment. He kept a diary of his impressions during three days of excitement on the island.

"I was one among many who had more important things to do at the far flung tracking stations and ships around the globe. We were keeping tabs on astronaut John Glenn as he orbited the Earth, sending back vital and new information about space flight in those early days. There was also a small but well-staffed Mercury hospital on the island as astronaut John Glenn would be brought there after orbiting the Earth and splashing down in nearby waters.

"We could see his heart beat on our equipment from the signals coming from outer space but unfortunately we were not able to hear him talk as it would have interfered with our own communications and operations. Later we listened to news reports and taped broadcasts on the Voice of America radio station so we didn't miss too much. Col. Glenn arrived at Grand Turk Island about 9:30 p.m., but I wasn't down at the air strip to greet him; I was asleep in bed, as I'd put in a very long day. We had arrived at the telemetry site many hours ahead of the scheduled launch to calibrate our equipment as well as participate in the flight readiness checkout of the whole Project Mercury tracking system around the world."

The following morning the aircraft carrier USS *Randolph* and two other smaller Navy ships were close by the island, where *Friendship 7* was brought ashore. "It was taken to the airstrip and loaded aboard a MATS (Military Air Transport System) aircraft to be shipped to Patrick Air Force Base, Florida. Later on in the day, three more astronauts flew in here; they were Shepard, Grissom and Schirra. As they came in and were walking to the mess hall for chow, I was able to get their pictures. I also took pictures of the space capsule loading operation at the dock and also at the airstrip. Col. Glenn stayed in the Mercury hospital while the other three astronauts bunked in the barracks – their quarters are about four doors down the hallway from me. Later on in the evening there was a party down at the Conch Club for Col. Glenn and everyone was invited. The astronauts showed up and were swamped by autograph seekers. I got my share of autographs too."

On Thursday, 22 February, a number of press conferences were held outside the hospital and the island was crawling with reporters from all over the world. "I attended a few of the press conferences to learn what was going on – it was the only way I could learn anything. I did work half a day though, but when I wasn't working I stayed around the Mercury hospital and my persistence paid off as I got more pictures of two additional astronauts who had arrived. Later on, all of the astronauts except Cooper posed for the photographers and TV men. I was right there with them snapping away. Most of the day the astronauts stayed in the Mercury hospital for the debriefing of Col. Glenn and for the final phase of his physical checkup."

At 4:30 a.m. on Friday morning, Vice President Lyndon Johnson arrived on the island in a C-140 JetStar. As Frasketi recalls, the arrival was about an hour earlier than expected. "He had breakfast at the mess hall with all the VIPs and the astronauts. A co-worker came and woke me, telling me that the Vice President was on the base, but we arrived at the mess hall a bit late to get any inside photographs. When all the VIPs came outside I was on hand for a photograph session. The Vice President and Col. Glenn were very friendly and posed for all of us. Johnson surprised us all by saying, 'You've been taking my picture all morning; why don't you men get around me and we'll have a picture taken together.'

"So we did with Col. Glenn and a few Secret Service men too. Mostly we were all base employees and amateur photographers as the professional photographers had already caught a plane to Patrick Air Force Base so as to be there when the Vice President and Col. Glenn arrived. While the VIPs took a tour of the Mercury hospital we all went to the airport to await their arrival. At the airport waiting area there was a large crowd of local townspeople; I would say about 200 natives and a few British dignitaries. When the Vice President arrived with Col. Glenn and the other VIPs, there was much handshaking and Vice President Johnson stole the show from Astronaut Glenn since he was a well-known and experienced hand-shaker. The crowd closed in on him wanting to shake his hand.

Vice President Lyndon Johnson and John Glenn pose for the group photo. Joe Frasketi is kneeling at the front. (Photo: Courtesy Joe Frasketi)

But Col. Glenn was still the hero as the crowd cheered together and hip-hip-hurrah'ed him. Then all the VIPs got aboard the JetStar and flew off to the U.S.A. Peace and quiet now prevails at Grand Turk Island once again. Now when Grand Turk is mentioned to someone in the States it will be recognized along with Astronaut Glenn.

"I was able to obtain the group photograph taken of us with Vice President and Col. Glenn and it is one of my prized possessions. I was disappointed to find out later, in reading local Florida newspapers, that Grand Turk Island was only mentioned briefly as a return point for the big news of Astronaut John Glenn's historic space flight. So the one big chance that this little island had to be spotlighted in a major news event was passed by."[24]

REFERENCES

1. *All Hands* (The Bureau of Naval Personnel Information Bulletin) issue No. 543, April 1962, pg. 3
2. Don Harter email correspondence with Colin Burgess, 18 November 2014
3. Carpenter, S., Cooper, Jr. L, Glenn, Jr., J., Grissom, V., Schirra, Jr., W., Shepard, Jr., A., and Slayton, D., *We Seven*, Simon and Schuster Inc., New York, NY, 1962
4. Jerry McConnell, JO1, *All Hands* (The Bureau of Naval Personnel Information Bulletin, No. 543, April 1962, pg.5
5. Carpenter, S., Cooper, Jr. L, Glenn, Jr., J., Grissom, V., Schirra, Jr., W., Shepard, Jr., A., and Slayton, D., *We Seven*, Simon and Schuster Inc., New York, NY, 1962
6. Jerry McConnell, JO1, *All Hands* (The Bureau of Naval Personnel Information Bulletin, No. 543, April 1962, pg.5
7. Peter H. Milliken, *Youngstown Vindicator* (Ohio) newspaper article, "An American Hero," published 12 February 2012

8. Carpenter, S., Cooper, Jr. L, Glenn, Jr., J., Grissom, V., Schirra, Jr., W., Shepard, Jr., A., and Slayton, D., *We Seven*, Simon and Schuster Inc., New York, NY, 1962

9. J.K. Herman, "Recollections of Project Mercury: a 35th Anniversary," *NAVY Medicine* magazine, Vol. 87, No. 3, May-June 1996

10. Carpenter, S., Cooper, Jr. L, Glenn, Jr., J., Grissom, V., Schirra, Jr., W., Shepard, Jr., A., and Slayton, D., *We Seven*, Simon and Schuster Inc., New York, NY, 1962

11. *Ibid*

12. Jerry McConnell, JO1, *All Hands* (The Bureau of Naval Personnel Information Bulletin, No. 543, April 1962, pg.5

13. Carpenter, S., Cooper, Jr. L, Glenn, Jr., J., Grissom, V., Schirra, Jr., W., Shepard, Jr., A., and Slayton, D., *We Seven*, Simon and Schuster Inc., New York, NY, 1962

14. Eugene Wolfe email correspondence with Colin Burgess, 18 September 2014

15. Richard Pomfrey email correspondence with Colin Burgess, 15–18 September 2014

16. Dean Conger email correspondence with Colin Burgess, 26 October 2014

17. Tom Hanko email correspondence with Colin Burgess, 18 September 2014

18. Robert Frangenberg email correspondence with Colin Burgess, 13 September–10 October 2014

19. Robert Bell email correspondence with Colin Burgess, 14 October 2014

20. John Glenn with Nick Taylor, *John Glenn: A Memoir*, Bantam Books, New York, NY, 1999

21. *Ibid*

22. Scott Carpenter telephone interview with Colin Burgess, 18 December 2002

23. *New York Times* newspaper, 24 February 1962

24. Joseph J. Frasketi, Jr., "The Grand Turk Island Connection with the Project Mercury/Glenn Flight," Astro Space Stamp Society journal *Orbit*, October 1998

8

Back home a hero

There is a little known story that McDonnell's pad leader Guenter Wendt kept to himself for many years, concerning fifty-two $1 bills that went into space aboard *Friendship 7*. Glenn had autographed the bills as mementos for employees of the McDonnell Douglas Aircraft Company who had worked closely with him on the development of the Mercury spacecraft.

"Somebody … and it was probably somebody who didn't get one of the dollar bills, complained about it," noted Wendt. "So we had somebody from Congress come down, I forget his name, and he got ahold of me and said, 'Do you realize you could've jeopardized the whole space mission with all those dollar bills floating around in the capsule?'" Wendt then asked the congressman if he knew exactly *how* they sent those dollar bills up. When there was no response, he explained in detail how they had taken the fifty-two dollar bills, intertwined them, and rolled them up like a thin noodle. Then they stuck them inside thermal shrink tubing, taking all the air out, so that it was a quarter of an inch thick. Tops. To doubly ensure that all the air was out, the tubing was placed inside an altitude chamber, after which it spent two days under 100 percent oxygen. A work order was then processed to install the tubing in the spacecraft with a bundle of other tubing, completely out of the way. "So you see, congressman," Wendt pointed out, "there weren't a bunch of dollar bills floating around inside the capsule." The congressman had to admit he hadn't realized the care that had gone into this little operation.[1]

A JOYFUL REUNION

On Friday, 23 February 1962, Vice President Lyndon Johnson flew down to Grand Turk Island in a Lockheed C-140 JetStar executive airplane. He was there to accompany John Glenn back to Florida, where the famed Marine's epic journey had begun three days earlier, and where he was now free to re-join his family.

On arrival at Patrick Air Force Base, south of Cape Canaveral, the Vice President was the first to disembark, shaking hands with a number of dignitaries. Then Glenn strode down the airplane steps with a happy smile and swooped into the arms of his family, who were there along with his parents John and Clara and Annie's parents, the Castors.

© Springer International Publishing Switzerland 2015
C. Burgess, *Friendship 7*, Springer Praxis Books, DOI 10.1007/978-3-319-15654-5_8

The tears flowed, and then he presented Annie with a small gold globe he had carried with him on his space flight.

Nearby, behind a short barricade, stood a crowd comprised of hundreds of well-wishers, who cheered, waved and applauded as he acknowledged their presence. A microphone had been set up, and when all was in readiness the Vice President stepped up to address those in attendance. The crowd became respectfully silent, then Johnson said, "It is a great pleasure to welcome home a great pioneer of history."

A playful Guenter Wendt "sympathizes" with John Glenn on one of the launch scrub days. (Photo: NASA)

Carolyn, John, Annie and David Glenn with Vice President Johnson at Patrick Air Force Base. (Photo: NASA/Ralph Morse)

Lyndon Johnson's speech was quite brief, but in it he mentioned that he had received the congratulations of Soviet Premier Nikita Khrushchev on the successful completion of the Mercury flight, and suggested that one day the two nations might cooperate in the peaceful exploration of outer space. With that he turned and addressed Glenn with a smile, and said, "So, you see, Colonel, you have done something that two presidents have been unable to do."

In response, Glenn said he was glad that everything had turned out so well, stressing that Tuesday's flight had been a team effort and that he regarded himself as just a representative of that team.

With those formalities at an end the Glenns joined the Vice President in his limousine as the lead vehicle in a twelve-car motorcade, plus camera and television trucks, ten press buses, and a battery of motorcycle policemen for the fifteen-mile trip to the Cape, where President Kennedy would attend an official welcoming ceremony. The cavalcade left Patrick Air Force Base right on 8:00 a.m. through what was normally a remote and desolate area. But it was far from quiet and tranquil that day, since tens of thousands of people lined the route to welcome home the nation's newest hero. As the open-top cars passed through Cocoa Beach, massive crowds cheered the astronaut, his family, the Vice President and the other astronauts; their applause and shouts of joy and congratulations mingled with and somewhat drowned out the patriotic sounds of the local high school band. Thousands of flags, bunting and signs hung from every vantage point through the town.

Eventually the motorcade reached the south gate of the Atlantic Missile Range where Glenn, in a most unnecessary gesture, showed his identification card to the smiling guards. There were more crowds along the roads inside, as hundreds of the 20,000 workers at the Cape had taken time off to see the astronaut go by.

At 10:22 a.m. *Air Force One* touched down on the Cape's airstrip, and soon Glenn was shaking hands with the President, before taking him on a guided limousine tour of the space base. Following this, everyone moved on to Hangar S, where a platform had been erected and the heat-stained *Friendship 7* was already prominently on display. It was here that the official ceremonies would take place.

CEREMONY AT THE CAPE

After making a few opening remarks, the President awarded NASA's Distinguished Service Medal to Robert Gilruth, in charge of Project Mercury. He then asked Glenn to step forward. Before he presented the beaming astronaut with his medal, Kennedy extolled the work Glenn had performed on the history-making flight three days earlier.

"Seventeen years ago, a group of Marines put the American flag on Mount Suribachi, Iwo Jima, so it's very appropriate that today we decorate Colonel Glenn of the United States Marine Corps, and also realize that in the not-too-distant future a Marine or a Navy man or an Air Force man will put the American flag on the Moon.

Excited crowds greet the Glenn family and Vice President Johnson in Cocoa Beach. (Photos: Associated Press)

The scene outside of Hangar S. Foreground is a mockup of a Mercury capsule; *Friendship 7* is to the right of photo. (Photo: NASA)

"I present this citation: The President of the United States takes pleasure in awarding the National Aeronautics and Space Administration Distinguished Service Medal to Lieutenant Colonel John H. Glenn, Jr., United States Marine Corps, for services set forth in the following:

"For exceptionally meritorious service to the Government of the United States in a duty of great responsibility as the first American astronaut to perform orbital flight. Lieutenant Colonel Glenn's orbital flight on February 20, 1962, made an outstanding contribution to the advancement of human knowledge, of space technology, and in demonstration of man's capabilities in space flight.

"His performance was marked by his great professional skill, his skill as a test pilot, his unflinching courage and his extraordinary ability to perform most difficult tasks under conditions of great physical stress and personal danger. His performance in fulfillment of this most dangerous assignment reflects the highest credit upon himself and the United States."

At this, President Kennedy pinned the medal on Glenn's chest, saying, "Colonel, we appreciate what you have done."[2]

After Glenn had made a short, well-received speech of thanks to everyone who had been a part of his flight, NASA Administrator James Webb declared the formalities at an end. At this time John Glenn and the President strolled over to the *Friendship 7* spacecraft, where the astronaut spent some time explaining different aspects of his space flight to

John Glenn shows his wife and President Kennedy his *Friendship 7* spacecraft. (Photo: NASA)

Kennedy, who listened intently. Following this, Glenn was able to show the spacecraft to his family. Then they all adjourned for lunch. Afterwards, Glenn made his way to a large, specially erected tent, where for more than an hour he answered questions during a packed press conference.

THE PRESS CONFERENCE

"Well, I know you are probably anxious to hear about what things were like on the flight," Glenn began, "so we might as well get started on that. I can make a few brief remarks concerning some of the major phases of it and then we can field some questions from it and then we will have some specific questions.

"First off, I – it was quite a day. I don't know what you can say about a day when you see four beautiful sunsets in one day. It is pretty interesting. Three in orbit and one on the surface when I was back aboard the ship. This is a little unusual, I think.

"We were asked some questions before the flight about how the delay affected our readiness and although I didn't read some of the write-ups, I understand there were some considerably gross misconceptions put out about how the delays affected our state of readiness. I think that Scott Carpenter put it into probably the best light when he said I was the most fortunate one around with all this excitement going on and all the business around the Cape area and all the preparations because all I had to do at that time was sit back and just hone the edge a little more and keep in good shape. I think that put it about in its proper light.

"I certainly did not experience any of the terrible subliminal impressions or hallucinations that I was supposed to be receiving from all of this pre-flight activity and cancellations. We've had our holds, of course, on previous launch attempts.

"On the day of the 20th, that morning, the weather started to clear and we could see the big blue holes coming up and I thought we were in a 'go' condition at that time because the booster was ready and everyone sort of gets 'go' fever when you get all conditions set right. I felt that was – I think everyone was excited that morning.

"The powered flight phase of the booster worked perfectly. Obviously it did from the fact that we came out within a very few feet per second of the velocity that we wanted and we were shooting of course for velocities a little over 25,000 feet per second. And when you come out within just a few feet per second at that speed, this is very accurate control. It was a good insertion. I think the best words I just about heard in my life were: 'You have a seven orbit capability' from Al Shepard here when he gave the 'go' to me from the communicator's position. That was a very welcome sound.

"I don't want to go into a lot of details on the launch – I don't think they're necessary. I think all that has been pretty well covered from what I have seen in the papers. As to our after-insertion – as to the zero g's – this, of course, is one of the big things we've wondered about through the early phases of the program, wondering what effect zero g's would have. This put us into a new environment that we have had no experience or limited experience before in airplanes. We extended this realm of weightlessness, of course, during the Redstone shots that Al and Gus had and we're extending it farther here and I'm very happy to report that there [were] no ill-effects I got from zero-g.

"It was very pleasant, as a matter of fact, and I had no tendency to particularly over-reach switches or have any trouble with the controls as a result of zero g's – indeed, a very, very pleasant experience. Someone told me last night after I had been talking so enthusiastically about this that I was an addict to it and I think I probably am. It's a wonderful feeling.

"We thought that over a period of time we might feel some ill-effects from it in the way of nausea or disorientation but there were none of these affects whatsoever. No problem keeping oriented under zero-g. I tried different head movements to see whether with certain head movements and combinations of head movements we would get any discomfort or any nausea. We did not experience any of these at all – and these wound up with some rather violent head movements in roll, pitch and yaw to see if we were going to induce any abnormal feelings. We did not. It was a very pleasant experience all the way through. Some of the things that happened under zero-g has sort of demonstrated how fast, though, a human being adapts to this situation.

"I recall last night, when we were discussing some of these things, the fact that I had had this little camera in my hand and when I had taken a picture and I wanted to do something with the switch immediately, it just seemed natural at that time, after I had been weightless for, I guess, a half-hour or forty-five minutes, I had acclimated to this rapidly enough, and it just seemed perfectly natural, rather than to put the camera away, I just put it out in midair and let go of it and went ahead with the switch position and reached back to the camera and went on with the work. I remember thinking afterward that we were treating this pretty blasé at the time here, but the point I'm making is that we all adapt very, very rapidly to these new situations, much more rapidly than those of us who have been training for this for three years would have believed possible. But it seemed just the natural thing to do. If you were just letting go of something for a minute, you'd just put it there and let go of it and go back to it when you get time.

"Under this zero-g condition, of course, it would be easy to lose different items, have them get away and get out of reach some place. As far as I knew there was only one item that got lost and that was a little can of film that I got out and was going to put in the camera. And I let it slip out of my fingers and went to grab for it and instead of clamping on to it I batted it and it went sailing off behind the instrument panel, and I think that was the last I saw of it. I don't know where it went.

"Eating experiments – we had planned to do some of that. We'd tried this before on some of the weightless flights in the airplanes, simulating weightlessness for a minute or so, and it hadn't appeared to be any big problem. However, we wanted to try it. And I did. I had one tube of food that was – that I squeezed into my mouth out of a tube – and this presented no problems swallowing or getting it down at all. I think the only restriction probably to food will be that it not be of a particularly crumbly nature like crumbly cookies that have little particles to them that might break off and you wouldn't be able to get all these back unless you had a butterfly net of some kind, I guess. As long as the food is solid, you can hold on to it and get it into your mouth, and from that point on there appears to be no problem. It's all positive action. Your tongue forces it back in the throat and you swallow normally and it's all positive – a positive displacement machine all the way through.

"Speed sensations – I've been asked repeatedly since the flight about the sensation of speed. Speed, of course, is relative. If you're in a complete – if you have nothing to refer to, you could be going almost any speed and you wouldn't have any sensation of it. I think the nearest I can come to it is – most of you have flown in jet airplanes at 30,000 feet or so and I think the feeling of speed is about the same as flying in an airliner such as that and looking down at clouds, say maybe at a height of 10,000 feet below you. You're at 30 say. I think the speed at which the ground goes out from under you – I mean just the sensation of speed you have at orbital altitude would be pretty close to that same speed. It's difficult to describe the speed sensation when you have nothing to relate it to. I think that's probably the closest that we can come to it.

"It was very noticeable, over a good part of our track, there were clouds of one type or another. They obscured some of the areas we wanted to look at. Part of the western United States, down across northern Mexico, large – almost the whole Pacific – was covered by one type cloud or another. First, coming down around Australia, Australia was always in the dark so we never could see Australia although we could see lights from the city of Perth. Cities around Perth, too.

"The sunset was probably the most impressive thing you would see in orbital flight. These are very brilliantly colored hues and the colors stretch way out from the Sun to the horizon. The horizon stays light for, I would estimate, some four or five minutes after sunset, which I found rather surprising because I thought the darkness would come on much more rapidly than that. I thought it would be just a matter of maybe a minute after the Sun went down. But apparently there is quite a bit of light curving around through the atmosphere that keeps the horizon very visible for a period of some five minutes or so.

"And this would bring us around to a sunrise. This turned into a pretty interesting area each time around, I think as most of you are aware. At first light of sunrise – the first sunrise I came to – I was still faced back toward the direction which I had come from with normal orbit attitude and just as the first rays of the Sun came up onto the capsule, I glanced back down inside to check some instruments and do something, and when I glanced back out, my initial reaction was looking out into a complete star field. The capsule had probably gone up while I wasn't looking out the window and I was looking into nothing but a new star field, but this wasn't the case, because a lot of little things that I thought were stars, were actually a bright, bluish-green, about the size and intensity of a firefly on a real dark night. And these little particles that were outside the capsule were – I would estimate – some six to ten feet apart, and there were literally thousands of them. As far as I could look off to each side I could see them. I could see them back along the path.

"Later on I turned around so that I was facing the direction from which they appeared to be coming and although, in that direction, toward the bright sunlight of the dawn most of them disappeared, you still could see a few coming toward the capsule. They appeared to have even distribution on each side of the capsule. I thought of two things that they might be initially. One was the cloud of needles that the Air Force put up some time ago and that appears to have disappeared somewhere. I thought that we had suddenly found them again. But they didn't look like that – they didn't look like they had any length to them at all. The other thing I thought was perhaps that – as we use our hydrogen peroxide jets – the hydrogen peroxide decomposes into steam and oxygen and comes out under high heat pressure. And I thought perhaps this water vapor was turning into little snowflakes and the luminous light from the Sun was causing them to fluoresce for some reason or other. But, when I would work the thrusters on the capsule, this did not appear to be causing any snowflakes or anything like it at all.

"So all I can say about these is that I saw them from about the first light of the Sun to a period of three and a half, or four minutes. The time period made close observation of them possible. Occasionally, one of them would come drifting by the capsule window very close in the side from the Sun and at that time they would look like very small white particles. And they would vary in size from maybe pinhead size to something say about three-eighths of an inch in diameter, or that order.

"There are numerous things that some of the people are thinking about looking into but I have no theory myself except that we observed them, we saw them on all three orbits, about the same length of time on each sunrise. They were very luminous – a yellowish-green color and, as George Ruff our psychiatrist, listened to this, he said, 'What did they say, John?'

"We had some problems during the flight. The automatic stabilization and control system was causing some difficulty. It didn't appear to be correcting up the way it should. I was able to use the manual control during that period and this appeared to – it didn't cause any trouble at all. It seemed very natural to take over manual control after all of the trainer work we've done. Trainer simulations, incidentally, were very, very close to the orbital control situation. This was particularly true in the fly-by-wire mode where we control through the automatic thrusters. When we started having this problem at the end of the first orbit, I was largely on manual control from there on until the end of the flight.

"The other observations that we had planned to make had to go by the board while I tried to work out the control system problems, and I spent most of the last two orbits working on this and making some other observations, largely concentrating about 90 percent of the time on the control system.

"Ground – as most of you probably know, the telemetry receivers had picked up an impulse that I possibly had a loose heat shield. And for that reason it was deemed advisable to keep the heat shield in place during re-entry so that it would go ahead and burn off and by that time we would be in a high enough aerodynamic field to keep the heat shield in place, in case it was, in fact, loose. This made [for a] pretty spectacular re-entry from the capsule standpoint because as the retro-package – as I retained the retro-package and entered into the first part of the high heat area of the re-entry, the straps on the retro-package broke loose and I felt a bump on the capsule and thought that the retro-package had jettisoned as it was supposed to do. Apparently, this was not true, but I thought so at the time. As I went on into the higher heat this glow picked up. Outside the capsule was sort of a bright orange glow outside the window.

"It became apparent that something was tearing up on the heat shield end of the capsule because there were large pieces – anywhere from pieces as big as the end of your finger to pieces probably seven or eight inches in diameter – were breaking off and falling off the edge of the capsule and coming up back past the window, very brightly. You could see the fire and the glow from them as they would come back up past the window.

'Well, this was obviously the retro-package burning up and breaking off as we knew it would if it had been retained. I thought at that time, however, that the retro-package had already been jettisoned. So there were some moments of doubt whether the heat shield had been damaged and whether it might be tearing itself up and this – this could have been a bad day all way around if this had been the case. But it was very spectacular looking out into this orange glow outside the window, bright orange glow, and seeing these big flaming chunks go back along the flight path.

"I understand there was some misconception also when we referred to this term blackout during re-entry, and I was questioned aboard ship as to how I had survived the blackout and when I regained consciousness, and a few things like this, and this was – when we referred to blackout during re-entry, a communications blackout is what we were referring to in this case. And we probably should re-term this as a 'com-loss' or something like that, so that we don't confuse it with a blackout of the astronaut involved. This is due to high ionization around the capsule and makes it impossible for the radio frequencies that we're using to get in and out from the capsule.

"G's during re-entry got up to about 8. [The] parachute functioned in completely normal fashion and that's probably the prettiest little old sight that you ever saw in your life – to look out the window and see that parachute. I remember Al Shepard making some comments about that after his flight and how pretty that parachute was and I concur. That's just about as pretty as anything you can see at that point.

"The capsule was pretty warm after re-entry – very hot, in fact – and it was getting very warm inside. I was sweating rather profusely when I came down and when I landed, so I remained as still as I could in the capsule, trying to add as little heat load to it as I possibly could. The destroyer *Noa* was close by and had me spotted and came over and made an excellent pickup of the capsule. It came aboard and at that time I had been heated up enough and had been sweating very profusely long enough that I decided after taking the panel – the right side of the panel – down, preparing the egress out through the top, I decided at that time, since we were already on deck, that it was hardly worth the effort and I should get on

Glenn speaking at the press conference, flanked by Robert Gilruth, James Webb, and Col. John Powers. (Photo: NASA)

out and be comfortable, so I did. We blew the hatch on the side, after checking to see that everyone was clear outside and came on out through the side hatch. I was not in such bad shape at that time that I could not have gotten out through the top if I had had to do it. I could still have come out through the top. But at that time I was hot. I'd been sweating for a long period of time and it seemed like the thing to do to get on out of there at the time.

"I think, perhaps, the difficulty that we had and the action I had to take may actually have been a blessing in disguise in one way, because it showed when we had to, that a man can take over control of the various systems and operate them manually and still know what he's doing and not have any ill-effects in this regard from zero-g. And probably, if sufficient study of our records from this flight warrants it, we probably can go on some flights with considerably less automation and less complexity, we hope, as a result of some of the things that we learned on this flight."

Having given his overview of the mission, Glenn invited questions from reporters.

Question: "Colonel Glenn, what did the stars and the Earth look like up there?"

Answer: "Well, stars – I was a little bit surprised, I think I maybe expected to see the stars in greater quantity and greater numbers than I had seen them before. The nearest thing I could compare it to and compare it to during debriefing was the – I think – if you've been out in the desert on a very clear brilliant night when there's no Moon up and the stars just seem to jump out at you, that's just about the way the stars look. We get some light reduction because of the light having to come through the window of the capsule, of course, and apparently this reduction in light may also approximate the losses that we normally get on a very clear night through the atmosphere. So I didn't feel that I could see any greater number of stars. They were not blinking. They came through very clear and straight, shining light. There was no flickering on and off. But other than that, it looked very similar to looking at the sky on a very, very clear night in the desert."

Question: "John, there were a number of experiments you were not able to make because, of course, you were busy flying the spacecraft. Could you give us a general idea of some of those that you were not able to perform this time, with a reflection here as to perhaps what we might expect from Deke Slayton's ride?"

Answer: "Well, there were several that we had wanted to look at. We wanted to take some infra-red pictures for the weather people. We wanted to take some ultraviolet pictures around on the dark side. We had some eye checks with some small instruments that we wanted to make. There were several things of this nature – I was going to eat some more, I never got around to that after our difficulties started a little bit."

Question: "Could you tell us some of the things you saw on Earth during the course of your flight while you were up there? Is there anything now that stands out in your mind as the most spectacular thing that you saw on the surface of the Earth?"

Answer: "Well, number one, it's all very spectacular from up there. You can see a tremendous distance. You're up above the atmosphere. You see this little horizon band of very brilliant blue color even on the day side. It would be difficult to pin down any one thing as being more spectacular than others. Certainly one of the most beautiful things is to be on the dark side with the full Moon out and see this coming off the clouds down below. And then running over and seeing the very sharp demarcation line where the stars keep coming down below the horizon. It's interesting to note that your sunsets and the stars moving down behind the horizon occur at approximately eighteen times their normal speed. This makes for a pretty speedy sunset."

Question: "John, you mentioned in the debriefing – made some observations about how you could see the Earth's surface – being able to tell the difference between land and sea."

Answer: "Yes, you can see difference patterns in the ocean currents, like the Gulf Stream, for instance. You can see the changing colors there. You can see on the Earth's – one area that I could observe very clearly was the area northwest of El Paso. And that's an area where there's a lot of desert with a big irrigated area that comes down a valley northwest of El Paso and that stood out very much. You can see the squares at the irrigated areas from this distance. I wouldn't guarantee that I was seeing the smallest irrigated squares that they have – each individual patch – but the larger irrigated squares probably had major dikes around them. I could see these very clearly."

Question: "Could you tell me what [was] the highest temperature during the flight?"

Answer: "I think the onboard cabin temperature, about the highest I observed, was about 105 [degrees Fahrenheit]. We're checking telemetry records and will probably have more accurate information on that later on."

Question: "Did you notice any difference in the accelerative effects on the body as opposed to the effects on the body at the end of the flight as a result of this period of weightlessness?"

Answer: "I noticed no effects at all. I felt just the same on re-entry as our runs on the centrifuge, where we were practicing for instance."

Question: "Colonel, how large a view did you have at any one time? For example, could you see all of Florida and more at one glance?"

Answer: "Yes, you can – well, when I was well up the east coast here once, I could see back across Florida and I think that I could see the Mississippi delta at that time. It was pretty clear along the gulf coast. Your view to the horizon is about – approaches 900 miles."

Question: "What was the noise level inside the capsule in orbit?"

Answer: "The noise level in the capsule was very similar to what it is like on the pad out here. You have the whine of the inverters, the gyros, the valves, the hiss of the oxygen in the helmet. It's not a loud intensity. I don't know what the decibel level of it is exactly, but it didn't appear to be any different in orbit than it is out here when we're running checks on the pad."

Following the presentation ceremony, John Glenn gives a double thumbs-up as he tours Cape Canaveral with President Kennedy. (Photo: NASA/KSC)

On a wet and dismal day in Washington, D.C., John Glenn's motorcade passes thousands of well-wishers on his way to the Capitol building. (Photo: United Press International)

Following Glenn's Q&A session, NASA Administrator James Webb hinted that Glenn and the other astronauts may become more active in non-space activities. He said the space program "has a lot to do with the United States' image." He made these comments after Glenn was asked whether he might go out of the country to help explain America's space program.

Once the press conference was at an end, John Glenn and his family flew out for a quiet, well-deserved weekend at the Naval Base in Key West, where they could simply be a family once again, walk on the beach, do a little boating and swimming, laze around and soak up some sunshine. It also gave Glenn time to prepare for another major event in his life, when he would participate in another, much larger motorcade in the nation's capital on Monday, followed by the tremendous honor of addressing a Joint Meeting of Congress.

"I worked on my speech, and on Monday morning we flew to the airport at West Palm Beach to meet the President for the trip back to Washington on *Air Force One* .... Somebody had typed my speech the night before, and as we flew north I asked President Kennedy to review it. He looked it over, said it was fine, and gave it back to me, and I put it in an inside pocket of my suit."[3]

A SPEECH BEFORE CONGRESS

It was raining for their arrival. Following a brief reception in the President's office in the White House, Glenn and his family were escorted to the open limousine for a motorcade parade along Pennsylvania Avenue to the Capitol. The route was packed with hundreds of thousands of eager well-wishers who had braved the dismal conditions and rain to welcome the national hero and his family, along with most of the other Mercury astronauts.

Dressed in overcoats and gloves, John, Annie and Lyndon Johnson sat on top of the back seat, while Lyn and David occupied jump seats just forward of them.

As the motorcade made its way to the Capitol, the House chamber was rapidly filling with people keen to hear the historic address. Over 1,000 spectators jammed the galleries which normally seated 740, while the House floor, meant to accommodate 448, somehow managed to seat over 600 congressmen, ministers, charges d'affairs of foreign governments, and other government officials.[4]

"It was a moment of intense gravity to me," Glenn recalled of standing at a lectern ready to give his address. "Through the standing applause, I was conscious that the opportunity to address [a Joint Meeting of] Congress was rare, a privilege reserved for royalty and heads of state. All the same, I felt at ease. I felt I was there not just as a test-pilot-turned-astronaut who happened to have made America's first orbital space flight, but as a patriotic American with something to say about the space program and its embodiment of our nation's persistent quest to expand our knowledge and press forward our frontiers."[5]

The following is a transcript of the speech given by John Glenn to the Joint Meeting of Congress on 26 February 1962.

Mr. Speaker, Mr. President, Members of Congress.

I am only too aware of the tremendous honor that is being shown to us at this Joint Meeting of the Congress today. When I think of past meetings that involved heads of state and equally notable persons, I can only say I am most humble to know that you consider our efforts to be in the same class.

This has been a great experience for all of us present and for all Americans, of course, and I am certainly glad to see that pride in our country and its accomplishments is not a thing of the past.

I still get a hard-to-define feeling inside when the flag goes by – and I know that all of you do, too. Today as we rode up Pennsylvania Avenue from the White House and saw the tremendous outpouring of feeling on the part of so many thousands of people I got this same feeling all over again. Let us hope that none of us ever loses it.

The flight of *Friendship 7* on February 20 involved much more than one man in the spacecraft in orbit. I would like to have my parents stand up, please. My wife's mother and Dr. Castor. My son and daughter, David and Carolyn. And the real rock in my family, my wife Annie.

There are many more people, of course, involved in our flight in *Friendship 7*; many more things, of course, involved, as well as people. There was the vision of Congress that established this national program of space exploration. Beyond that, many thousands of people were involved, civilian contractors and many subcontractors in many different fields; many elements – civilian, civil service, and military, all blending their efforts toward a common goal.

To even attempt to give proper credit to all the individuals on this team effort would be impossible. But let me say that I have never seen a more sincere, dedicated, and hardworking group of people in my life.

From the original vision of the Congress to consummation of this orbital flight has been just over three years. This, in itself, states eloquently the case for the hard work and devotion of the entire Mercury team. This has not been just another job. It has been

As Vice President Johnson listens, John Glenn delivers his speech to a Joint Meeting of Congress, 26 February 1962. (Photo: United Press International)

a dedicated labor such as I have not seen before. It has involved a crosscut of American endeavor with many different disciplines cooperating toward a common objective.

Friendship 7 is just a beginning, a successful experiment. It is another plateau in our step-by-step program of increasingly ambitious flights. The earlier flights of Alan Shepard and Gus Grissom were stepping stones toward *Friendship 7*. My flight in the *Friendship 7* spacecraft will, in turn, provide additional information for use in striving toward future flights that some of the other gentlemen you see here will take part in.

Scott Carpenter here, who was my backup on this flight; Walt Schirra, Deke Slayton, and one missing member, who is still on his way back from Australia, where he was on the tracking station, Gordon Cooper. A lot of direction is necessary for a project such

Among those attending the historic speech were five of Glenn's fellow Mercury astronauts (seen just above his head). Seated behind Wally Schirra is Robert Gilruth, Director of Project Mercury. (Photos: United Press International)

as this, and the Director of Project Mercury since its inception has been Dr. Robert Gilruth, who certainly deserves a hand here.

I have been trying to introduce Walt Williams. I do not see him here. There he is up in the corner. And the Associate Director of Mercury who was in the unenviable position of being Operational Director. He is a character, no matter how you look at him. He says 'hold the countdown' occasionally, and one thing and another.

With all this experience we have had so far, where does this leave us? These are the building blocks upon which we shall build much more ambitious and more productive portions of the program.

As was to be expected, not everything worked perfectly on my flight. We may well need to make changes – and these will be tried out on subsequent three-orbit flights later this year, to be followed by eighteen-orbit twenty-four-hour missions.

Beyond that, we look forward to Project Gemini – a two-man orbital vehicle with greatly increased capability for advanced experiments. There will be additional rendez-vous experiments in space, technical and scientific observations – then, Apollo orbital, circumlunar and finally, lunar landing flights.

What did we learn from the *Friendship 7* flight that will help us to attain these objectives? Some specific items have already been covered briefly in the news reports. And I think it is of more than passing interest to all of us that information attained from these flights is readily available to all nations of the world.

The launch itself was conducted openly and with the news media representatives from around the world in attendance. Complete information is released as it is evaluated and validated. This is certainly in sharp contrast with similar programs conducted elsewhere in the world and elevates the peaceful intent of our program.

Data from the *Friendship 7* flight is still being analyzed. Certainly, much more information will be added to our storehouse of knowledge. But these things we know. The Mercury spacecraft and systems design concepts are sound and have now been verified during manned flight. We also proved that man can operate intelligently in space and can adapt rapidly to this new environment.

Zero-g or weightlessness – at least for this period of time – appears to be no problem. As a matter of fact, lack of gravity is a rather fascinating thing. Objects within the cockpit can be parked in midair. For example, at one time during the flight, I was using a hand-held camera. Another system needed attention; so it seemed quite natural to let go of the camera, take care of the other chore in the spacecraft, then reach out, grasp the camera and go back about my business. It is a real fascinating feeling, needless to say.

There seemed to be little sensation of speed although the craft was traveling at about five miles per second – a speed that I, too, find difficult to comprehend.

In addition to closely monitoring onboard systems, we were able to make numerous outside observations.

The view from that altitude defies description. The horizon colors are brilliant and sunsets are spectacular. It is hard to beat a day in which you are permitted the luxury of seeing four sunsets.

I think after all of our talk of space, this morning coming up from Florida on the plane with President Kennedy, we had the opportunity to meet Mrs. Kennedy and Caroline before we took off. I think Caroline really cut us down to size and put us back in the proper position. She looked up, upon being introduced, and said, "Where is the monkey?" And I did not get a banana pellet on the whole ride.

Our efforts today and what we have done so far are but small building blocks in a huge pyramid to come. But questions are sometimes raised regarding the immediate payoffs from our efforts. What benefits are we gaining from the money spent? The real benefits we probably cannot even detail. They are probably not even known to man today. But exploration and the pursuit of knowledge have always paid dividends in the long run – usually far greater than anything expected at the outset. Experimenters with common, green mold, little dreamed what effect their discovery of penicillin would have.

The story has been told of Disraeli, Prime Minister of England at the time, visiting the laboratory of Faraday, one of the early experimenters with basic electrical principles. After viewing various demonstrations of electrical phenomena, Disraeli asked, "But of what possible use is it?" Faraday replied, "Mister Prime Minister, what good is a baby?"

That is the stage of development in our program today – in its infancy. And it indicates a much broader potential impact, of course, than even the discovery of electricity did. We are just probing the surface of the greatest advancements in man's knowledge of his surroundings that has ever been made, I feel. There are benefits to science across the board. Any major effort such as this results in research by so many different specialties that it is hard to even envision the benefits that will accrue in many fields.

Knowledge begets knowledge. The more I see, the more impressed I am – not with how much we know – but with how tremendous the areas are that are as yet unexplored.

Exploration, knowledge, and achievement are good only insofar as we apply them to our future actions. Progress never stops. We are now on the verge of a new era, I feel.

Today, I know that I seem to be standing alone on this great platform – just as I seemed to be alone in the cockpit of the *Friendship 7* spacecraft. But I am not. There were with me then – and with me now – thousands of Americans and many hundreds of citizens of many countries around the world who contributed to this truly international undertaking voluntarily and in a spirit of cooperation and understanding.

On behalf of all those people, I would like to express my and their heartfelt thanks for the honors you have bestowed upon us here today.

We are all proud to have been privileged to be part of this effort, to represent our country as we have. As our knowledge of the universe in which we live increases, may God grant us the wisdom and guidance to use it wisely.[6]

There would be more motorcades and parades to come for John Glenn. According to later police estimates, around four million people turned out on a freezing cold day in New York City on 1 March to cheer and applaud as Glenn and the other astronauts slowly rolled by in their open limousines. It was easily the greatest traditional ticker-tape reception since that accorded to the revered American pioneer aviator, Col. Charles Lindbergh. Another post-parade estimate put the amount of ticker tape and other associated paper later cleaned up by the city's sanitation department at some 3,500 tons.

Massive crowds lined the streets of New York as Glenn's motorcade slowly made its way through a veritable rainstorm of paper and tickertape that almost hid the limousines from view. (Photos: United Press International)

"DEAR COLONEL GLENN"

Two days later a smaller but none the less enthusiastic crowd of around 75,000 people turned out to see Glenn during a parade through the streets of his home town of New Concord, Ohio.

At a homecoming ceremony held in the Patton Hall of Muskingum College, the college president, Dr. Bob Montgomery, announced that the college gymnasium would henceforth be known as the John Glenn Physical Education Center. But this was just the beginning of the hometown honors that would be heaped on the astronaut and his wife. Plans were revealed to build a new school north of New Concord that was to be called the John Glenn High School. Governor Michael DiSalle also declared that Route 40 from Cambridge to Zanesville would now be known as John Glenn Highway, and that Upper Bloomfield Road would be renamed Friendship Drive, thereby honoring the spacecraft that had carried him around the world.

In his speech, an overwhelmed Glenn thanked everyone for turning out that day and shared a few memories of growing up in the town. "From swimming down over the hill in the college lake to outer space is a pretty big jump," he said. "It's good to be an American, and it's good to be home."[7]

When he finally returned to Houston he was astonished to find three large United States mailbags of congratulatory letters and autograph requests awaiting him on the living room floor of the new home that he and Annie had built, but his amazement at this outpouring of universal pride in his achievement was completely overshadowed when a postman came to their door and announced that the Post Office was holding nearly a truckload of additional mailbags. "I realized the three bags that friends had placed in our living room were just a token of what actually had arrived." He and Annie opened and read what he would later describe as just the first installment of what was to become a real avalanche of mail.[8]

There was even more mail awaiting him in his office at the newly established Manned Spacecraft Center. "It soon became obvious we could not begin to open and sort all of the mail personally. Fortunately, Mr. Steve Grillo, one of our friends in the Administrative Services Division of NASA Headquarters, foresaw what a problem this was to become and

John Glenn surveys the mountain of mail awaiting him following his Mercury space flight. (Photo: NASA)

took steps to organize and handle the mail. Thanks to his efforts and those of his assistant, Mrs. Amelia Leukhart, we were able to cope with this responsibility."[9]

At the time of the writing of this book, more than five decades after his history-making Mercury flight, John Glenn daily continues to receive mail from across the United States and all around the world, and to his great credit he still sets time aside to respond to every letter.

REFERENCES

1. Peter Kerasotis, *Florida Today* newspaper article, "Pad Leader recalls final moments of '62 liftoff," issue 29 October 1998, pg. 12A
2. Lt. Col. Philip N. Pierce, USMC and Karl Schuon, *John H. Glenn: Astronaut*, George G. Harrap & Co. Ltd., London, U.K., 1962, pp. 166–167
3. John Glenn with Nick Taylor, *John Glenn: A Memoir*, Bantam Books, New York, NY, 1999
4. Lt. Col. Philip N. Pierce, USMC and Karl Schuon, *John H. Glenn: Astronaut*, George G. Harrap & Co. Ltd., London, U.K., 1962, pg. 175
5. John Glenn with Nick Taylor, *John Glenn: A Memoir*, Bantam Books, New York, NY, 1999
6. Lt. Col. Philip N. Pierce, USMC and Karl Schuon, *John H. Glenn: Astronaut*, George G. Harrap & Co. Ltd., London, U.K., 1962, pp. 178–184
7. Ken Kettlewell, *Our Town: New Concord, Ohio, The Birthplace of John Glenn*, Express Press, Lima, Ohio, 2001
8. John H. Glenn, Jr., P.S., *I Listened to Your Heartbeat: Letters to John Glenn*, World Book Encyclopedia Science Service, Inc., Houston, TX, 1964
9. *Ibid*

9

Epilogue: Beyond the Mercury program

Following his history-making flight aboard *Friendship 7* in February 1962, 40-year-old John Glenn felt that he could still contribute his talents to another space flight or two. However, a feeling of unease soon began to replace those of enthusiasm and optimism. Occasionally, he would drop into the office of Robert Gilruth to ask when he might be placed back in the crew rotation, but his boss seemed evasive and would not commit to an answer. Instead, he would offer explanations such as, "Headquarters doesn't want you to go back up – at least not yet."

Initially, Glenn accepted this, but as the months dragged by and his astronaut colleagues began picking up mission assignments in the two-man Gemini program he became more and more dispirited with this attitude. Having grown frustrated by this seeming lack of interest in having him fly another mission, he grudgingly resigned from NASA on 16 January 1964. It was time to think about moving on.

PILOTING, PUBLIC RELATIONS AND POLITICS

Glenn was philosophical when later asked if he would have liked to fly to the Moon on Project Apollo. "I was forty when I made my flight," he mused, "and I thought I would be fifty before I got a chance for a lunar flight. To be the oldest astronaut in a permanent training status did not seem to be an ideal career, so I looked elsewhere."[1]

Navy psychologist Robert Voas was the astronaut training officer for the Mercury Program, and he recalls this frustrating time for the famed astronaut:

"John, of course, after his flight, he's such a charismatic and effective spokesperson that he was essentially kidnapped by Headquarters almost immediately and used in PR [Public Relations] work. This was not really his choice, though of all the seven, John was the one who was most adept at that and most comfortable doing it.

"NASA was constantly working on budgets, particularly at that time we were beginning to plan both Gemini and the lunar flights, and so it really helped having a very charismatic, highly popular national hero to go visit congressmen and speak at [Joint Meetings of Congress] and be sent around the world and so on. John was also the oldest of the group.

© Springer International Publishing Switzerland 2015
C. Burgess, *Friendship 7*, Springer Praxis Books, DOI 10.1007/978-3-319-15654-5_9

On 8 February 1963, almost a year after his historic space flight, John Glenn was interviewed by Associated Press correspondent Howard Benedict. (Photo: Associated Press/Ed Kolenovsky)

So I think he came to the feeling fairly quickly that he was very unlikely to get another flight, and he had always been interested in government, and so he became interested in the possibility of running for the [United States] Senate in Ohio, and I worked with him on plans for that.

"He was encouraged to do that by the Kennedys. He and Jack Kennedy took a very strong liking to each other, and the Kennedys had John and Annie [Glenn] out to their home on more than one occasion and to the White House."[2]

Glenn always harbored suspicions that he had deliberately been held back from making another space flight. "It was only years later that I read in a book that [President] Kennedy had passed the word that he didn't want me to go back up. I don't know if he was afraid of the political fallout if I got killed, but by the time I found out, he had been dead for some time, so I never got to discuss it with him."[3]

As one of the most high-profile and popular American identities of that time, Glenn was increasingly being wooed by a number of companies, foundations and political parties.

As Bob Voas further recalls: "I came up to Washington for a day to work at the NASA Headquarters and then went on to Ohio. When I got into the taxi after my day at NASA, they were telling about the assassination of Kennedy in Dallas. So I arrived in Ohio to meet with the Democratic leaders there on the day that Kennedy was assassinated, which was not the best time to start a political candidacy. But John did decide to go ahead, and in January of 1964, even though Kennedy was no longer president – Johnson was – he started the campaign there in Ohio. Then I resigned from NASA to go with him and work with him on the campaign. So that's how I came to leave NASA."[4]

According to Glenn's longtime friend and campaign aide Warren Baltimore, "I don't think he would have retired from the space program if he thought he had any chance of going to the Moon. But, in the absence of that, he believed the Senate represented a way for him to continue in a role of public service."[5]

Mercury astronaut Wally Schirra was always known for telling things the way they were, and he related John Glenn's entry into the political arena with the eloquent but over-the-top pilot that he sat with when the seven Mercury astronauts were introduced at a packed press conference back in April 1959: "John Glenn craved the publicity. I think even John would admit that. When he went into politics, that became pretty obvious! We all saw that he knew how to do public relations from that original press conference. We weren't prepared for that at all – we were all looking over, thinking, what is this guy saying?"[6]

Another fellow Mercury astronaut, Alan Shepard, agreed that Glenn had a way of swaying the masses, and an obsession with setting a good example to the young: "John always acts as if he were being watched by an army of Boy Scouts or children. Even when he is scratching his nose or peeing."[7]

SPACEMAN AND SENATOR

Early indications were that once Glenn had turned to politics he would stand an excellent chance of becoming a United States Senator on his first try for office. Still stunned by the assassination of President Kennedy the year before, he settled on making a run for the U.S. Senate in 1965 as a Democrat in his native state of Ohio. Then an Earthbound slip changed everything.

On the morning of 26 February 1964, two years after his orbital flight, John Glenn had just finished shaving in his bathroom when he tried to close the sliding doors on his wall cabinet. They wouldn't budge. He removed one of the mirrored panels to determine what had caused it to stick in the track, but as it came out, the scatter rug that he was standing on slipped on the tiles. Still holding the mirrored panel he fell backwards and hit the back of his head on the side of the bathtub. "The mirror flew from my grip and crashed to the floor," he later recalled. "Flying pieces of glass cut my hands and face. I didn't pass out, but I was in a state of shock." He received a severe bruise and concussion and was taken to hospital for x-rays. Two weeks later he was still having dizzy spells and could not maneuver well. "The whole thing is completely incongruous," he later suggested. "I had never been in a hospital before except for checks. I've never been injured. When I think of all the crazy things I've been through – all the sports, two wars, 149 combat missions, getting my plane hit twelve times by ack-ack, all the training for the space flight and the three orbits themselves. In all that, and the worst I ever got was a knuckle skinned getting out of my *Friendship 7* capsule. And now this!"[8]

While the blow from his fall had given him mild concussion, it also caused Glenn to suffer from labyrinthitis, or damage to the mechanism of the inner ear affecting balance. He was taken to the aerospace medical clinic in San Antonio, where he spent the next two months immobile in bed, suffering bouts of vertigo and vomiting. He couldn't move his head as much as an inch without being gripped by severe nausea. It was a full six months before he could even stand upright without suffering any ill effects.

In preparing for life outside of NASA and the military, Glenn also put in his notice of resignation from the U.S. Marines, but this was turned down until he was completely fit.

On 27 October 1964, a delighted John Glenn became a full colonel in the U.S. Marine Corps when he received his commission from President Johnson in a White House Rose Garden ceremony. (Photo: United Press International)

In October 1964, despite his intention to retire from the service, John Glenn was promoted to full colonel in the U.S. Marines. Four months earlier, he had written to the Marine Corps asking that he not be considered for promotion as it was his intention to resign from the corps after 22 years of service. He felt that promotion should be predicated on the future value of an officer, and as there could only be a certain number of colonels in the corps – at that time 605 – he felt he might be holding back someone who deserved the promotion more than him. However the Secretary of the Navy, Paul Nitze, ordered that the number of colonels in the Marine Corps be increased by one to 606, and this was authorized by President Johnson, who said that "Glenn was being proposed for promotion despite his letter because Secretary of the Navy Nitze and the Marine Corps desired to recognize his many accomplishments while in the service of his country." An extra promotion was added to the list, thereby ensuring that no one missed out. When informed of the President's actions, Glenn said the nomination "comes as a surprise," and he was "very gratified that this has occurred."[9] Thus, it was as a full colonel that Glenn would officially retire from the Marine Corps on 1 January 1965.

Earlier, in late March 1964, acting on the advice of his doctors, the bed-ridden astronaut withdrew from the Senate race. "Doctors warned me that without proper convalescence the condition would worsen and perhaps become permanent." That year he decided – for a time – to enter the quieter life of private enterprise, and was appointed to the board of Royal Crown Cola, later becoming president of the company. He also served on the boards

Glenn with his official certificate of retirement from the U.S. Marine Corps. (Photo: NASA)

Senator Robert Kennedy and his wife Ethel with John Glenn at a fundraising dinner in December 1967 at the Plaza Hotel, New York. (Photo: United Press International)

of a variety of other corporations and made investments in hotel developments. "I had to re-plan my whole life," he commented. "It has not been easy."[10]

Glenn's health slowly improved and he even took a jet flying refresher course, but he was still picking up the pieces of a life shattered by the accidental fall. He used the time to further his business career.

Another tragedy came his way in Los Angeles on the afternoon of 6 June 1968. That day the Glenns were on the campaign trail with his old friend, mentor and presidential front-runner Bobby Kennedy. They were at the Ambassador Hotel the night Kennedy was gunned down and critically wounded. The next few hours were filled with chaos for everyone as doctors tried to save his life. Kennedy's wife Ethel asked the Glenns to take her youngest children home to Massachusetts, but within hours of arriving they were told that Bobby had lost the struggle and passed away. It was a distraught John Glenn who had to sit down with the Kennedy children and tell them their father had died. "It was the hardest thing I've ever had to do in my life," he said.[11]

In January 1970, Glenn announced his candidacy once again as a Democrat for the U.S. Senate, but despite his high profile and obvious charisma he lost to Cleveland millionaire Howard M. Metzenbaum, who ran a powerful, heavy spending campaign. Politically and privately devastated, Glenn shook off the defeat and made a determination to work harder and try again. Four years later, in November 1974, he finally defeated Metzenbaum in the primary and was elected to the U.S. Senate.

Metzenbaum had made the miscalculation of mounting a personal attack on Glenn's work ethic, saying that his opponent had spent the majority of his life on the public payroll. The final nail in his political coffin came when he challenged Glenn by saying, "How can you run for the Senate when you've never held a job?"

Glenn's stinging impromptu rebuke at a press conference at the City Club of Cleveland clearly demonstrated the fire that still burned within him, and it was John Glenn at his very finest:

"I served 23 years in the United States Marine Corps. I fought through two wars. I flew 149 missions. My plane was hit by anti-aircraft fire on 12 different occasions.

"I was in the space program. It wasn't my checkbook; it was my life that was on the line. This was not a nine-to-five job where I took time to take the daily cash receipts to the bank.

"I ask you to go with me as I went the other day to a Veterans Hospital and look those men with their mangled bodies in the eye and tell them they didn't hold a job. You go with me to any Gold Star mother, and you look her in the eye and tell her that her son didn't hold a job. You go with me to the space program, and you go as I have gone to the widows and the orphans of Ed White, and Gus Grissom, and Roger Chaffee, and you look those kids in the eye and tell them that their dad didn't hold a job. You go with me on Memorial Day coming up, and you stand in Arlington National Cemetery – where I have more friends than I like to remember – and you watch those waving flags, and you stand there, and you think about this nation, and you tell me that those people didn't have a job.

"I tell you, Howard Metzenbaum, you should be on your knees every day of your life thanking God that there were some men who held a job. And they required a dedication to purpose and a love of country and a dedication to duty that was more important than life itself. And their self-sacrifice is what has made this country possible …

"I *have* held a job, Howard!"

What followed from the City Club crowd was a standing ovation. The following Tuesday, Glenn won the Democratic primary by 91,000 votes.[12]

Senator Glenn went on to serve four terms in Congress, during which he held key posts on several committees, including the Committee on Government Affairs. Outspoken on many issues, he campaigned for more funds for space exploration, science and education. Among other accomplishments, he was chief author of the 1978 Non-Proliferation Act, he served as chairman of the Senate Government Affairs Committee from 1978 until 1995, and he sat on the Foreign Relations and Armed Services committees and the Special Committee on Aging.

Glenn was also a contender for the Democratic Vice Presidential nomination in 1976, but his keynote address at the Democratic National Convention failed to impress the delegates. The nomination instead went to veteran politician Walter Mondale.

In 1984 Glenn decided to run in the Democratic primaries as a presidential candidate. He obviously had some advantage and national recognition based on his accomplishments as an American space hero, and in a timely development a blockbuster film featuring his flight into space, *The Right Stuff*, was released almost concurrently with the 1984 primaries. However, he steadfastly refused to take political advantage of the hype associated with this movie. He maintained it would be dishonorable to use the resultant publicity to further his campaign.

Senator John Glenn and his wife Annie. (Photo: Ohio State University)

Senator and future Vice President Joseph Biden in discussion with Senator John Glenn in Washington, D.C., October 1981. (Photo: Associated Press)

Sadly, he failed to ignite sufficient interest in the party faithful and was placed a decidedly poor fifth place with a four percent showing in the Iowa Caucuses, which mortally wounded any serious chance he had at winning. When further results easily demonstrated that he stood no chance against his Democratic rivals Walter Mondale and Gary Hart for the opportunity to challenge the immensely popular Ronald Reagan for the presidency, he conceded defeat and withdrew from the race, dejected, and with massive personal debts.

He had lost his chance to run for the presidency of the United States, but for John Glenn there was one more great challenge left to him, and this time, fourteen years later, he would succeed.

BACK INTO ORBIT

It was a day John Glenn had dreamed of since achieving his epic three orbits of the Earth in 1962, but one he never imagined would ever come around. In October 1998, after thirty-six years, history came full circle when the 77-year-old astronaut flew into space again aboard Space Shuttle *Discovery*, this time on a nine-day mission that, once it had been announced, quickly enthralled and delighted the vast majority of the American people.

In 1995, while still serving as a member of the Senate Special Committee on Aging, Glenn decided to embark on a bold plan that might see him fly into space once again. NASA had compiled vast dossiers on his medical status back in 1962; so why not conduct experiments on a space shuttle flight in order to compare those results with tests carried out when he was nearly double the age he had been during Project Mercury? By the summer of 1996 he had amassed sufficient material to present his case to NASA on the phenomenon of aging, and arranged a meeting with Daniel Goldin, the agency's Administrator.

"I told him there are 34 million Americans over sixty-five, and that's due to triple in the next fifty years," Glenn later told *Time* magazine. "And I told him someone ought to look into this."[13]

Goldin was a pragmatist; he could see the medical merit in Glenn's argument, and also understood the massive boost such a flight of an American hero would mean to the agency. But first it would be necessary to garner supporting information to justify giving the elderly Senator a shuttle seat, since, as he told Glenn, "We've got no open seats just for rides."[14]

Goldin and Glenn had to present a convincing case in the face of concerns from a number of people about the actual merits of such a proposal. Eventually the barriers were overcome as the consultant doctors and scientists concluded that, while somewhat tenuous, there was sufficient science involved to make such a flight, and further determined that Glenn's health and motives for making the flight were quite sound. Meanwhile, on 20 February 1997 – the 35th anniversary of his Mercury flight – the former astronaut announced he was planning on retiring from the U.S. Senate.

It was a long and nervous wait, but on 15 January 1998, Glenn received a phone call from Goldin advising that he would be returning to space, and a public announcement would be made the next day. As a consequence of being assigned to mission STS-95 aboard Space Shuttle *Discovery*, and the deep training this involved, Glenn made his retirement from the U.S. Senate official. He then began preparing for the challenge of making his second space flight, this time as a payload specialist.

On 16 January 1998, a press conference was held to announce that John Glenn would be returning to space as a crewmember aboard Space Shuttle *Discovery*. Glenn is shown at the press conference alongside NASA Administrator Dan Goldin. (Photo: NASA)

A formal portrait of Annie Glenn with her astronaut husband. (Photo: NASA)

In fact Annie Glenn was initially against the idea, after his twenty-four years in the U.S. Senate. "I was looking forward to having him as my own, because I had given him to our government for fifty-five years. Then he was named to go back up again, and I wasn't 100 percent behind it at all then."[15] But those reservations quickly melted away in the ensuing months of intense flight training, and the entire Glenn family gave him their unqualified support for what he was doing. The involvement of the legendary Mercury astronaut would make it one of the most publicized, closely perused, and highly antici-pated shuttle flights in many years.

NEARING LAUNCH

On the eve of the shuttle launch, with a quarter of a million people expected to be in the vicinity of the Cape to witness the lift-off, Glenn's surviving fellow Mercury astronauts participated in an open-air news conference at the U.S. Astronaut Hall of Fame nearby the

Kennedy Space Center. They knew they would face questions about the supposed aging studies purpose of their colleague's flight, and they decided to tackle the issue head-on.

"We might as well deal with the controversial nature of this flight first off," 73-year-old Scott Carpenter told the gathered reporters. "In my view, John's return to space has been a marvelous thing for the country. It's been a marvelous thing for the space program and NASA. It is an honest search for new truths in the scientific discipline of human physiology."

He added that just like the Mercury missions they had flown in the early 1960s, Glenn's flight would be the first step in an emerging scientific field to link the debilitating effects of weightlessness with the aging process on Earth.

"Every time you try to do something new in any scientific field, you find naysayers. They are an occupational hazard to people who want to do forward-looking work. I think this is a great opportunity for the country to revitalize its interest in the space program. John will bring back some good new information, so everybody wins from this flight, including John Glenn, who is going to have the time of his life."[16]

Dr. William Douglas, once the personal physician to the Mercury astronauts, was also there, and he was equally delighted to see John Glenn fly into space once again.

"If I were emperor of the world, I would let every single Mercury astronaut take a flight in the shuttle as a reward from a grateful nation," Douglas stated. "You see, they did something that was really hazardous back then, flying on rockets that had a nasty habit of exploding. So what should we do now? Well, they had the glory back then. Now let them have the fun."[17]

The crew of Space Shuttle *Discovery*, STS-95. Top row from left: Scott Parazynski (mission specialist), Stephen Robinson (mission specialist), Chiaki Mukai (payload specialist), Pedro Duque (mission specialist) and John Glenn (payload specialist). Bottom row: Steven Lindsey (pilot) and Curtis Brown (commander). (Photo: NASA)

Harvey Wichman, the director of the Wreospace Psychology Laboratory at Claremont McKenna College in California, best summed up the euphoria surrounding Glenn's shuttle flight in a piece he wrote for the *Florida Today* newspaper:

"After a distinguished career in the U.S. Senate, and with the flame of the spirit of *Friendship 7* still burning passionately in his breast at age 77, he is 'gonna do it!' Will we citizens pay for it? You bet we will. This is our space program. Of course, we want to get good science from it. Of course, we aren't giving rides to just anyone who comes along. Glenn knows this. We will get some very good work out of him on this trip – he would have it no other way. But the space program that we people know and have created and paid for is not just about science and technology. It is also about spirit and people and aspirations and competence and thrills and risks and all those things that make our hearts beat faster and cause us to say things like, 'this is a great time to be alive.'

"To those who have spoken ill of both Glenn's flight and those in NASA who authorized it, I say, watch and be amazed by the interest in this flight, both in America and around the world. You have missed the point of what the space program is all about. Oh, yes, it is about science, engineering, and technology. But it is also much more than just that. It is about that indomitable American spirit that shone so brightly in the lift-off of *Friendship 7* and still burns unabated in the heart of its pilot at age 77. We all have that spirit and have it manifested for us in this flight of Glenn's. And it is worth every penny spent!

"So go for it, John Glenn! You got the right to go back into space the old-fashioned way. You earned it!"[18]

At 2:19 p.m. on the afternoon of 29 October 1998, *Discovery* ripped into the skies above the Kennedy Space Center from Launch Pad 39B on a mission scheduled to last nine days. There was a crew of seven astronauts aboard, including payload specialist John Glenn.

Shortly before lift-off, NASA had played for the crew a taped greeting from Mercury astronaut Scott Carpenter that was to be radioed to them after launch. His voice quivering slightly, Carpenter intoned, "Good luck, have a safe flight, and … once again, Godspeed, John Glenn," repeating his famous benediction for Glenn's first flight in 1962.

Medical research carried out during *Discovery*'s 134 orbits of the Earth included a battery of tests on John Glenn and his fellow payload specialist, Spanish-born Pedro Duque, to gather data on how weightlessness might affect a person's balance and perception, immune system response, bone and muscle density, metabolism and blood flow, and sleep.

Just after midday on 7 November, mission commander Curtis Brown brought *Discovery* down to a smooth touchdown back at the Kennedy Space Center after a highly successful 3.6 million-mile journey around the planet that lasted a total of 9 days, 19 hours, 54 minutes and 2 seconds.

All of the crew, including an elated 77-year-old space pioneer, were checked and found to be fit and well after a flight which, whatever its merits, had truly captured the imagination of the American people.

One old friend who had been at the Kennedy Space Center for the launch of John Glenn aboard *Discovery* was Dr. William Douglas (Lt. Col., USAF, ret.), a flight surgeon who had been assigned the personal physician for the Mercury astronauts, with responsibility for their medical care and flight preparation. Aged 76, Bill Douglas became ill while returning with his wife to their home in Albuquerque, New Mexico, and passed away from a viral infection on 15 November, eight days after the triumphant return of *Discovery*.

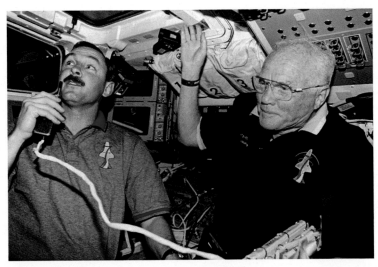

Back in orbit once again, payload specialist John Glenn floats free aboard *Discovery* along with mission commander Curtis Brown. (Photo: NASA)

John Glenn joined a formal photograph at the 2008 Ohio Astronaut Reunion. The All-Ohio group comprised of (top row, left to right): Mary Ellen Weber, Michael Foreman, Michael Gernhardt, Kevin Kregel, Don Thomas and Sunita Williams; (middle row): Nancy Currie, Terry Henricks, Ron Sega, Mark Brown, Greg Harbaugh, Tom Hennen and Carl Walz; (front row): Ken Cameron, Bob Springer, Neil Armstrong, John Glenn, Jim Lovell and Kathy Sullivan. (Photo: Don Thomas/NASA)

A LIFETIME OF ACHIEVEMENT

John Glenn has been the recipient of countless honors throughout his life and stellar careers with the U.S. Marines, NASA, the U.S. Senate, and elsewhere – as a result of which many highways, ships, schools, scholarships and numerous other institutions now bear his name.

In his lifetime he has been awarded the Distinguished Flying Cross on six occasions, in addition to the Air Medal with eighteen Clusters for his service during World War II and Korea. He also holds the Navy Unit Commendation for service in Korea, the Asiatic-Pacific Campaign Medal, the American Campaign Medal, the World War II Victory Medal, the China Service Medal, the National Defense Service Medal, the Korean Service Medal, the United Nations Service Medal, the Korean Presidential Unit Citation, the Navy's Astronaut Wings, the Marine Corps' Astronaut Medal, the NASA Distinguished Service Medal, and the Congressional Space Medal of Honor.

John Glenn, first American to orbit the Earth, was presented with one of his greatest awards at the White House on 29 May 2012 – the Presidential Medal of Freedom. (Photo: Courtesy MCT, The Lantern News/Media Website)

In 1976 he was enshrined in the National Aviation Hall of Fame, and in 1990 was inducted into the U.S. Astronaut Hall of Fame. Ten years later, he received the annual U.S. Senator John Heinz Award for Greatest Public Service by an Elected or Appointed Official, and in 2004, was awarded the Woodrow Wilson Award for Public Service by the Woodrow Wilson International Center for Scholars of the Smithsonian Institution. In 2009, he was awarded an Honorary LL.D from Williams College, and the following year received an Honorary Doctorate of Public Service from Ohio Northern University.

Among numerous other awards and decorations, John Glenn has also received the Congressional Gold Medal, the National Geographic Society's Hubbard Medal, and the Thomas D. White National Defense Award.

However one of the greatest honors accorded the beloved former astronaut and U.S. Senator occurred on 1 March 1999, when the Lewis Research Center in his home state of Ohio was officially renamed the NASA John H. Glenn Research Center at Lewis Field.[19]

As he wrote towards the end of his memoir published that same year: "I expect the future to bring new rewards and challenges. And as always, going forward requires touchstones in the past. My parents' legacy was honest hard work, sweat and dirt, effort and the grasp of opportunity. They believed in themselves and in their country, and they had faith in God. That legacy has been my guide through all the places I've been and all the things I've been fortunate enough to do, and I've tried to pass it along."[20]

REFERENCES

1. Ray Kerrigan, article, "Spaceman on way up again," Sydney *Daily Mirror* newspaper, issue 24 October 1964
2. Dr. Robert B. Voas interviewed by Summer Chick Bergen for NASA JSC Oral History program, Vienna, Virginia, 19 May 2002
3. Jeffrey Kluger, article, "Back to the Future," *Time* magazine, Vol. 152, No. 7, 17 August 1998, pg. 43
4. Dr. Robert B. Voas interviewed by Summer Chick Bergen for NASA JSC Oral History program, Vienna, Virginia, 19 May 2002
5. Jeff Lyttle, article, "John Glenn: An American Story," *Columbus Monthly* magazine, issue August 1998
6. Francis French, interview/article with Wally Schirra, "I worked with NASA, not for NASA," 22 February 2002. Available at: *http://www.collectspace.com/news/news-022202a.html*
7. Oriana Fallaci, *When the Sun Dies*, (English translation), Atheneum House, Inc., London, U.K., 1966
8. Sydney *Daily Mirror* newspaper article, "Glenn out of Senate Race," issue 26 March 1965
9. *Chicago Tribune* newspaper, "Johnson Asks Promotion of Ex-Astronaut," issue 30 September 1964
10. Anthony Syme, *The Astronauts*, Horwitz Publications, Inc., Sydney, Australia, 1965, pg. 54
11. Jeff Lyttle, article, "John Glenn: An American Story," *Columbus Monthly* magazine, issue August 1998
12. Mark Shields, *Lawrence Journal-World* (Lawrence, Kansas) newspaper article, "He Wasn't Elected President, But No Doubt, Glenn Held a Job," issue 22 February 1999, pg. 8B
13. Jeffrey Kluger, article, "An American Hero," *Time* magazine, issue Vol. 152, No. 7, 17 August 1998
14. *Ibid*
15. Robyn Suriano, *Florida Today* article, "Crew takes it easy as clock ticks down," issue 28 October 1998, pg. 1
16. Todd Halvorson, article "Astronauts cover launch for TV," *Florida Today* newspaper, issue 29 October 1998, Pg. 12A
17. *Ibid*

18. Harvey Wichman, article, "Seven, 77 are happy, lucky numbers for Glenn," *Florida Today* newspaper, issue 28 October 1998, pg. 15A
19. Wikipedia entry, "John Glenn," online at: *http://en.wikipedia.org/wiki/John_Glenn*
20. John Glenn with Nick Taylor, *John Glenn: A Memoir*, Bantam Books, New York, NY, 1999

Afterwords

On Friday, 9 October 1998, having learned of the imminent retirement of John Glenn from the U.S. Senate, his friend and fellow politician, Senator Carl Levin (Democrat, Michigan) asked to make a formal statement for the Congressional record.

"Mr. President, when the 105th Congress adjourns *sine die* in the next few days, the Senate will lose one of our nation's true heroes, and one of my personal heroes, Senator John H. Glenn, Jr. of Ohio. I rise today to pay tribute to this great American, a man I feel genuinely honored to call my friend.

"All of us old enough to remember John Glenn's flight into orbit around the Earth on February 20, 1962, aboard *Friendship 7* stand in awe of his courage and strength of character. But this enormous accomplishment followed on a distinguished record of heroism in battle as a Marine officer and pilot. He served his country in the Marine Corps for 23 years, including his heroic service in both World War II and the Korean conflict. And, in turn, his remarkable accomplishment in the history of space flight has been followed by an extraordinary Senate career over the past 24 years, as the only Ohio Senator in history to serve four consecutive terms.

"For the 20 years that I have been in the Senate, I have served side by side with John Glenn in both the Governmental Affairs Committee which he chaired for many years and now serves as Ranking Minority Member and the Armed Services Committee where he serves as the Ranking Minority Member of the Subcommittee on Airland Forces. More recently, I have served with John Glenn on the Senate Select Committee on Intelligence. This has given me a front row seat to watch one of the giants of the modern day U.S. Senate do the hard, grinding work of legislative accomplishment.

"Over the years, John Glenn has led the fight for efficiency in government, for giving the American people more bang for that tax 'buck.' He was the author of the Paperwork Reduction Act. He has worked to streamline federal purchasing procedures, and led the fight to create independent inspectors general in federal agencies. He was the point man in the Senate for the Clinton Administration's battle to reduce the size of the federal workforce to the lowest levels since the Kennedy Administration. He and I have fought side by side to block extreme efforts to gut regulatory safeguards in the name of reform and for the passage of a sensible approach to regulatory reform to restore confidence in government

© Springer International Publishing Switzerland 2015
C. Burgess, *Friendship 7*, Springer Praxis Books, DOI 10.1007/978-3-319-15654-5

regulations. Throughout his career, John Glenn has made himself an enemy of wasteful spending and bureaucracy, yet a friend of the dedicated federal worker.

"John Glenn has steadfastly served as a powerful advocate for veterans. He led the effort to bring the Veterans Administration up to cabinet level and to provide benefits to veterans of the Persian Gulf conflict. On the Armed Services Committee, John Glenn has brought his enormous credibility to bear time and again both in that Committee and on the Intelligence Committee on the side of needed programs and weapons and against wasteful and unnecessary ones like the B-2 bomber.

"Perhaps John Glenn's most important role, however, has been as the author of the Nuclear Non-Proliferation Act and as the Senate's leader in fighting the proliferation of nuclear weapons around the world. In this area, the Senate will sorely miss his clear vision, eloquent voice and consistent leadership.

"Mr. President, John Glenn, of course, has remained the strongest and most effective voice in the Senate for the nation's space program. Many of us will be on hand to watch the launch of his second NASA mission later this month, 31 years after the first. At age 77, John Glenn has volunteered to go back into space to test the effects of weightlessness on the aging process, and once again inspires our nation and sets an example for us all – an example of courage, character, sense of purpose, and, yes, adventure.

"No person I've known or know of has worn his heroism with greater humility. John Glenn is, to use a Yiddish word, a true mensch, a good and decent man.

"John Glenn and his beloved wife, Annie, are simply wonderful people. They, their children and grandchildren are the All-American family. My wife Barbara and I will keenly miss John and Annie Glenn as they leave the Senate family."

Another person to pay tribute to Glenn and his many achievements is retired U.S. Marine Corps Colonel Jack Lousma. He was selected as a NASA astronaut in April 1966. When asked, the veteran of two space missions was delighted to contribute a few words about the 'senior' Marine of the astronaut corps.

"United States Marine Corps Colonel John Glenn was the first American to orbit the 'Good Earth.' He strapped the Atlas rocket onto his back, lit the engines, and 'rode the fire' into space on February 20, 1962. He was also the first U.S. Marine Corps aviator to fly into space. Colonel Glenn's flight aboard his *Friendship 7* Mercury space capsule lasted nearly five hours and circled the Earth about three times.

"During the same hours Colonel Glenn was flying in space, I was merely a Marine First Lieutenant in the final phases of combat training for a one-year assignment in the Far East with a squadron flying the A-4 Skyhawk, a jet-attack aircraft used primarily for Close Air Support of Marine troops on the ground. The probable location of our overseas mission required completion of a Winter Survival Course that was conducted for about a week at a Marine Corps training facility appropriately named 'Pickle Meadows.' This course included prisoner-of-war training and mountain survival, as well as escape and evasion techniques in the event of an aircraft ejection over hostile territory.

"The training was very realistic, and I was dumped off in six feet of snow among towering pine trees at 9,000 feet in the mountains of California. My job was to stay warm and inconspicuous in a snow cave, built by piling snow on dead tree limbs over a cavity hollowed out in the snow. A 'friendly partisan' would periodically show up to ensure I was still alive. One day, the partisan passed by and stated, 'John Glenn is up.' A few hours later, the partisan reported, 'John Glenn is back.' Imagine the pride I felt knowing a fellow Marine I had never met, chosen for his professional performance and personal character, was the first American to orbit the Earth. Little did I know, then merely a First Lieutenant, that I would become the third active-duty Marine Corps aviator after Colonel Glenn and Major C.C. Williams to join NASA's astronaut roster. That night, I 'escaped and evaded' through a mountain snowstorm back to 'friendly territory,' but Colonel John Glenn's escapade in space was foremost in my mind.

"When I arrived at NASA in April of 1966, the astronaut ranks swelled in number from 30 to 49 in preparation for the Apollo landings on the Moon. Survival training in the California mountains now seemed far away. I had admired John's accomplishments and lifestyle from afar, and now my wife Gracia and I were personal friends with John and Annie Glenn. We occasionally crossed paths with them during his career as a U.S. Senator and, more recently, in various space-related events with other astronaut colleagues. Despite all the publicity and deserved accolades, John and Annie Glenn are among the most personable and humble celebrities in America."

Al Worden joined NASA in 1966, in the same astronaut group as Jack Lousma, and would serve as Command Module Pilot for the Apollo 15 mission in 1971. He recalls Glenn as: "A very personable man, as evidenced by his time as a senator. Politically motivated even when in the program, and he used that to his advantage by cultivating the friendship of JFK. He very cleverly parlayed his fame into the Senate. And I think the guys – like John Glenn, I think the challenge of being a senator in terms of its value to civilization is much greater than going to the Moon. The fact that he got there because he had been the first one to go into Earth orbit, I think that's what got him there, but I think once he was there – he can affect millions of lives by being a senator. He can only affect one going into orbit by himself."

In 2012, on the 50th anniversary of John Glenn's orbital flight aboard *Friendship 7*, the accolades and tributes were led by fellow Ohioan Neil Armstrong, the first man to walk on the Moon. At a celebratory gala event at Ohio State University, Armstrong was the surprise speaker who said that Glenn was "no ordinary pilot" and in the early 1960s when a need for leadership in the space program was required; he "literally rose to the occasion."

REFERENCES

1. Senator Carl Levin on the Congressional Record, 9 October 1998
2. Jack Lousma, email correspondence with Colin Burgess, 23–27 November 2014
3. Al Worden, interview with Francis French for *Falling to Earth: An Apollo 15 Astronaut's Journey to the Moon*, Smithsonian Books, Washington, DC, 2011
4. Unaccredited online article, "Neil Armstrong full of praise for John Glenn's Historic Flight." Available at *http://freestarcharts.com/index.php/19-news-and-current-events/62-neil-armstrong-full-of-praise-for-john-glenns-historic-flight*

Appendix 1

Deployment of the MA-6 recovery force

For the MA-6 mission, a task force of twenty-four Navy ships and over sixty aircraft, along with many supporting specialized units, was positioned from Cape Canaveral across the Atlantic to the Canary Islands. The organization responsible for this support was known as the Project Mercury Recovery Force, under the command of Rear Admiral John L. Chew, USN, Commander, Destroyer Flotilla Four.

MA-6 was the 20th recovery operation in which the Mercury Recovery Force had participated. The actual composition of the forces varied with each recovery, but was normally comprised of ships, aircraft, and Marine helicopters from the U.S. Atlantic Fleet, aircraft from the AFMTC (Air Force Missile Test Center), aircraft and pararescue teams of the Air Rescue Service, and Light Amphibious Resupply Craft (LARC) from the Army. Admiral Chew exercised overall control of the recovery force from the recovery room located next to the control room in the Mercury Control Center at Cape Canaveral.

The area off-shore to Bermuda was assigned to a recovery group under the command of Capt. Charles H. Morrison, Jr., Commander, Destroyer Squadron 24, embarked in the USS *Blandy* (DD-943). Units of this group were:

USS *Blandy*, USS *Cone* (DD-866), USS *Goodrich* (DD-831), USS *C. S. Sperry* (DD-697), USS *Observer* (MSO-461), USS *Exploit* (AM-440), USS *Recovery* (ARS-43), and four P2V aircraft from Patrol Squadron 18.

From Bermuda to approximately halfway across the Atlantic, a group under the command of Rear Adm. William Ellis, Commander, Carrier Division Two who flew his flag in the USS *Forrestal* (CVA-59), consisted of:

USS *Forrestal*, USS *Barry* (DD-933), USS *Stormes* (DD-780), USS *Norfolk* (DL-1), USS *Glennon* (DD-620), USS *Witex* (DD-848), four Lockheed WV aircraft from Airborne Early Warning Training Units Atlantic, four Martin P5M aircraft from Patrol Squadron 45, five P5M aircraft from Patrol Squadron 49, four Douglas SC-54 aircraft from the 55th Air Rescue Squadron, four Lockheed P2V aircraft from Patrol Squadron 18, and three Sikorsky HUS helicopters from Marine Air Group 26.

The area assigned the above group included the site selected for the spacecraft landing if it had been decided to terminate the flight after one orbit.

© Springer International Publishing Switzerland 2015
C. Burgess, *Friendship 7*, Springer Praxis Books, DOI 10.1007/978-3-319-15654-5

In the Eastern Atlantic area, a group under the command of Capt. Donald G. Dockum, USN, Commander, Destroyer Development Group Two, embarked in the USS *Hugh Purvis* (DD-709), was made up of:

USS *Chukawan* (AO-100), USS *Hugh Purvis*, USS *Brownson* (DD-868), USS *Sarsfield* (DD-837), four WV aircraft from Airborne Warning Squadron 44, and four Grumman SA-16 or Douglas SC-54 aircraft from the Air Rescue Service.

South of Bermuda, at a site selected for landing at the end of the second orbit, under the command of Capt. James H. Armstrong, was made up of:

USS *Antietam* (CVS-36), USS *Kenneth D. Bailey* (DD-713), USS *Turner* (DD-834), three P5M aircraft from Bermuda Patrol Unit, and three HUS helicopters from Marine Air Group 26.

At the end of the third orbit, about 200 miles northwest of San Juan, a group under the command of Rear Adm. Earl R. Eastwold, Commander, Carrier Division 16, embarked in the USS *Randolph* (CVS-15), comprised of:

USS *Randolph*, USS *Noa* (DD-841), USS *Stribling* (DD-867), six P2V aircraft from Patrol Squadron 16, two SA-16 and two SC-54 aircraft of the Air rescue Service, and three HUS helicopters from Marine Air Group 26.

The planned method of retrieval was by surface ship or helicopter. All of the deployed ships had conducted pick-up exercises with dummy spacecraft.[1]

Twenty top medical specialists of the U.S. Air Force, Navy, and Army were ready to rush to Glenn if they were needed, and each ship in the recovery force was staffed with a medical team and a surgeon. They were part of the overall recovery team that totaled 167 physicians, nurses, technicians and other experts in the recovery operation.[2]

REFERENCES

1. NASA *Space News Roundup*, "Mercury Recovery Force is Deployed," NASA Manned Spacecraft Center, Houston, TX, issue 21 February 1962, pg. 8
2. Donald Harter, email correspondence with Colin Burgess, 23 October 2014

Appendix 2

Sequence of events during the MA-6 flight

EVENT	PLANNED TIME hr:min:sec	ACTUAL TIME hr:min:sec
Booster engine cut-off (BECO)	00:02:11	00:02:10
Tower release	00:02:34	00:02:33
Escape rocket firing	00:02:34	00:02:33
Sustainer engine cut-off (SECO)	–	00:05:01
Tail-off complete	00:05:04	00:05:02
Spacecraft separation	00:05:04	00:05:04
Retrofire initiation	04:32:58	04:33:08
Retro (left) No. 1	04:32:58	04:33:08
Retro (bottom) No. 2	04:33:03	04:33:13
Retro (right) No. 3	04:33:08	04:33:18
Retro assembly jettison	04:33:58	–
0.05-g relay	04:43:53	04:43:31
Drogue parachute deployment	04:50:00	04:49:17
Main parachute deployment	04:50:36	04:50:11
Main parachute jettison (water impact)	04:55:22	04:55:23

© Springer International Publishing Switzerland 2015
C. Burgess, *Friendship 7*, Springer Praxis Books, DOI 10.1007/978-3-319-15654-5

Appendix 3

In his own words

The following was a guideline for John Glenn to follow in taping his shipboard debriefing ahead of a far more extensive debriefing on Grand Turk Island. Glenn was asked to read the questions and respond accordingly.

This debriefing kit contains the astronaut shipboard briefing outlines and the remote site debriefing questionnaire. In the execution of this debriefing, the debriefing team members or shipboard representatives will not interrupt the astronaut. The astronaut will read the question and answer on the voice recorders. Debriefing team members will take notes and ask all candid questions at the completion of the formal debriefing. This mode of operation is essential in order to obtain clean and transcribable records of the debriefing.

ASTRONAUT SHIPBOARD DEBRIEFING OUTLINE

1. What would you like to say first?
2. Briefly describe the highlights of your flight.
3. Evaluate your general condition at this time and during the flight.
4. What operational or capsule systems problem did you encounter?
5. What physiological effects did you notice?
6. Tell about the period of weightlessness.
7. Comment on your performance during the flight. Where did it satisfy you most? Dissatisfy you most?
8. What mental effects did you notice that have not been covered? In order to recapture some of the effects that might easily be lost with intervening experience, comment on the conditions in which you were:

   (a) Uncomfortable
   (b) Amused
   (c) Anxious

© Springer International Publishing Switzerland 2015
C. Burgess, *Friendship 7*, Springer Praxis Books, DOI 10.1007/978-3-319-15654-5

    (d) Fascinated
    (e) Elated
    (f) In pain
    (g) Preoccupied
    (h) Surprised
    (i) Bored
    (j) Distracted
    (k) Other.

9. Comment on each of the following if any phenomenon occurred that was not part of the flight plan:

    (a) Vibration
    (b) Noise
    (c) Sounds
    (d) Lighting
    (e) Relative movement
    (f) Orientation
    (g) Other sensory or perceptual experiences.

10. Describe the flight, using the appended sequence of flight events. (Pages 1–22 of MA-6-13 Flight Plan).

SUMMARY OF MA-6 ORBITAL FLIGHT BY JOHN H. GLENN, JR.

(Forenote: Within this overall NASA report there is a somewhat confusing amalgamation of at least two post-flight debriefing reports given by John Glenn, which results in a degree of repetition in descriptions and details).

There are many things that are so impressive, it's almost impossible to try and describe the sensations that I had during the flight. I think the thing that stands out more particularly than anything else right at the moment is the re-entry during the fireball. I left the shutters open specifically so I could watch it. It got to a brilliant orange color; it was never too blinding. The retropack was still aboard and shortly after re-entry began, it started to break up in big chunks. One of the straps came off and came around across the window. There were large flaming pieces of the retropack – I assume that's what they were – that broke off and came tumbling around the sides of the capsule. I could see them going on back behind me then and making little smoke trails. I could also see a long trail of what probably was ablation material ending in a small bright spot similar to that in the pictures out of the window taken during the MA-5 flight. I saw the same spot back there and I could see it move back and forth as the capsule oscillated slightly. Yes, I think the re-entry was probably the most impressive part of the flight.

PRE-LAUNCH

I thought in general, things went pretty smoothly during the pre-launch phase. I got all the information I really needed. The actual insertion itself went very smoothly. We've been through this a number of times and Joe Schmitt knows exactly which strap to pick up and when. He does a wonderful job of getting you plugged into the capsule. So there was no problem from that standpoint at all. The practice that we had had before I think stood us in pretty good stead for a smooth insertion because we find we're working as a team. When Joe reaches for a certain strap I just automatically turn my body to let him pull it out to a certain point and when he's tightening it up, I know just where to squeeze in a little bit to help him. I guess maybe it sounds rather peculiar but even something like that becomes sort of a team maneuver. We have been through this so many times that it was very easy and went very smoothly, I thought. We had some problems of course, such as the problem with the mike in the helmet that delayed us a little bit. I guess we have probably pushed those mikes up and down many thousands of times with no trouble and it had to pick that particular time to break, right in the middle of a count. It was the little fitting that slides up and down on the left mike that broke. Joe was able to change it rapidly. We had problems with the hatch bolt also, which delayed us for a little while. But this was no problem as far as I was concerned.

I was surprised at the leak rate we achieved. All they would read it as was below 500 and this is phenomenal. The best we have ever registered before on capsule 13 as I recall, was 610. So this was excellent. I think that the constant bleed was probably greater than the leak rate. Later on we did build up a pressure slightly above ambient in the capsule. The card storage for the star charts was not ideal at all, even for getting them in and out on the pad. We need some better way of stowing things. The map book storage just doesn't seem to be too good either; you shove that map book in and out of that case trying to review things and leafing through it is very awkward. If the pages are open, this adds to the difficulty. I don't know how we are going to improve it. I felt that it was unnecessary to go back over the switch positions a second and third time during the count. It seemed to me that the last time we go over the switches and set the fuses is completely redundant and unnecessary, a sort of WPA program [i.e. Works Project Administration, or providing jobs for the unemployed] designed to keep the astronaut busy. I was not particularly annoyed, I didn't have anything else better to do at the time; but to have something in there just to keep the astronaut busy isn't necessary.

POWERED FLIGHT

I think lift-off was just about as I had expected. The vibration information that [aeronautical performance engineer] Sig Sjoberg had obtained for us was very accurate. I could feel the engines fire up and when they fired, the whole bird shook, not violently but very solidly. There is no doubt about when lift-off occurs. I had thought the booster might lift-off so gently that there might be some doubt as to when you were actually moving; but there's no doubt about it, you know when you come off the pad. When the Atlas releases, there's an immediate surge – a gentle surge – that lets you know you're underway. Some vibration occurred immediately after lift-off for some 10 to 15 seconds. The roll to the correct azimuth was noticeable right after we lifted off. I had the little mirror pre-set to watch the ground. I glanced up right after lift-off and I could see the horizon going around. After about 10 or 15 seconds of flight, the vibration seemed to smooth out some, but not really as much as I'd anticipated. I thought we would go through a smooth flight period before we got to the high-Q area, but it never did smooth out completely. The vibration reduced somewhat but there was still quite a bit of vibration and shuddering that you could feel very distinctly. This continued up until we got into the high-Q area. You feel a lot more solid shudders and a lot more intense vibration. I think this started a little earlier and lasted a little longer than I thought it would from our pre-flight briefings. It is difficult to be sure whether you are feeling gimbal or whether it is only the general shaking which you sense. You can hear a roar all during this time; it's a dull muffled sound about the same intensity and type of sound as the roar we used on the centrifuge at Johnsville. That was a pretty good simulation of it. The vibration continued through the high-Q area for a minute and 15 or 20 seconds. Beyond that point it smoothed out very noticeably and I think I commented on this over the radio. However, the spacecraft never became completely vibration free; there were always some small vibrations that were noticeable.

The acceleration buildup was noticeable but it was not particularly bothersome. I believe I was still making voice reports or talking at maximum g, just before BECO.

The capsule came around and I believe overshot a little in its orbit attitude position in yaw. I lost sight of the booster but when the capsule corrected itself to the proper position, I could see the booster by leaning down a little and to the left and looking up a little. When we settled down into proper attitude the booster was right in the position that Carl Huss [NASA Flight Operations Division] had been predicting it would be, above the top of the window and a little to the left. Anytime I wanted to see it, I could lower my head and see it up in that position. Then it slowly drifted down to the window where I had a constant view of it for I guess probably some 6 to 8 minutes. The last time I saw it, it was probably a mile or so away from me and about a half a mile below my level in a slightly lower orbit. I looked for it later over Kano [Nigeria] where it was supposed to be directly below me but I couldn't see it in the periscope at all. There was no sensation of tumbling head over heels like those you experience on the centrifuge with its rotary motion. Before the flight, Scott [Carpenter] said he thought it would feel good to go in a straight line acceleration rather than just around in a circle. It did feel good; I knew I was accelerating with a purpose this time rather than just making circles on a wheel.

I think I was rather surprised at BECO. The BECO cut-off was not as sharp a line as I had anticipated it might be. It felt as though it ramped down a little over approximately half a second. BECO was very comfortable. I checked the accelerometer and it was indicating about one and one fourth to one and one half g, right where it should be. At that time you could feel the booster engines leaving. This creates some vibration as these come off. I think you could also hear them. I think you could feel the whole booster shake at that time a little. I don't recall any sensation of head over heels tumbling at this time. I did get just a little bit of this sensation at SECO later on but that, once again, was not as pronounced as I expected. I can remember having just a little feeling at SECO as though I had tilted forward, as though my head was going over a little toward my heels but it was very slight. It may be that the capsule actually did pitch down enough at SECO so that I felt the angular acceleration from this motion rather than just an illusion of movement. At BECO I felt the g reduction and then the slow buildup in acceleration again.

At BECO I saw out the window what looked like some smoke coming back by the capsule. I caught just a flash of this out the window. Before the flight, we had noted that at booster staging some smoke and fire did appear up in the area of the capsule. Though we had discussed this a number of times the first thing that entered my mind was that the tower had fired early. And I think I made a comment on this over the radio. Of course, it had not fired. It did jettison on schedule at 2:34 and I was all set to back it up at that time. I looked out the window for fire. I had my hand up on the pull ring waiting for it. I had checked the clock and got in sync with it. I was counting the seconds to 2:34 because I wanted to look out the window and watch it fire. I was looking right at the nozzles when they fired, just a big belch of smoke coming out. It wasn't a very bright light at all. The tower really accelerated from the capsule; it moves fast when you fire it. I watched it to a distance of approximately one half mile. It appeared to go straight. It didn't appear to be tumbling at all. I was just looking at the back end of it as it was going away in a straight line with the center of the window. This was a little surprising because the alignment of the tower is such that it should take off back over your right shoulder on a 120 degree course, relative to the spacecraft heading, so it shouldn't have stayed in the middle of the window.

At tower jettison, the spacecraft pitched down and provided my first real view of the horizon and clouds. I could see the horizon behind the tower as it jettisoned. It came down below the horizon maybe 3 or 4 degrees. I could see clouds out across the Atlantic. The horizon was just visible. After the tower fired, the spacecraft immediately started pitching up again, and I lost sight of the horizon. There was no marked roughness during this transition period to guidance. I don't recall any bumps or shudders of the capsule as a result of guidance starting. It was a smooth flight from there on. The pitch was about on nominal. I think I called out a couple of variations from the values the Capsule Communicator was reading. I don't recall the exact numbers of them now but there was only a slight variation. I remember making comment about the sky getting very black. When I looked out once before that, it had some color to it and it was very noticeable at this time that it was just black.

As we came up toward SECO, the acceleration built up. There was more vibration and, just before SECO, it was very noticeable. There was one sensation I was not expecting. At this time the tanks are getting empty, and apparently the booster becomes considerably more limber that it normally is. You have the sensation of being out on the end of a spring board. You feel a lot of oscillating motions, as if the nose of the booster were waving back and forth a little bit. The oscillations were not high in frequency. It is like being on the end of a big, long, loose spring, which is what you are, I guess. I remember wondering at the time how close these oscillations would get to our abort limits; obviously, they did not trigger the ASIS [Abort Sensing and Implementation System]. I recall I heard a little more noise coming up to SECO, too. I don't know whether this noise echoes up through the hollow tank. Right at SECO I had the very slight sensation of tumbling forward just a little bit. I don't know whether it was just a sensation due to cut-off or whether it was due to a natural pitch down that the capsule made at that point.

There is no doubt at all when the clamp ring fires. Boy, you can hear those rings fire! And I was surprised at the force that posigrade rockets had, too. I've always thought of the posigrades as being pretty insignificant little rockets. I thought that if you were not paying much attention you might miss the whole program on posigrade. But there is no doubt on posigrade fire. They boot you right on off the booster. The capsule damped turnaround in good shape. As I recall the spacecraft pitch came up a little bit as we turned around. I think maybe it pitched a little high coming around and then corrected back down again. As we came around, I could see the booster. I bet it wasn't 100 yards away. The nose of the booster was pointing off to north or northeast.

BOOSTER SIGHTING

Well, that was a real good sight. As it came to almost its proper position, the capsule was pitched up a little bit and here was the booster right in the middle of the window. It wasn't more than 100 to 150 yards away. That was some sight, this great big booster sitting right there, just that far away, and going away, thank goodness. It was sitting at an angle with the capsule into the booster pointed down maybe 30 degrees and the longitudinal axis of the booster angle maybe 45 degrees to my position so that it was pointed down and northeast. As the capsule corrected back, then settled – pitched down a little bit – into orbit attitude, the booster went out of sight at the top of the window. I could duck my head down though and look up and see the booster. I have to give Carl Huss credit. He drew up the chart on exactly where the booster would be and he couldn't have hit it better if he had been there, I don't think. I watched, on and off, for some 6 or 7 minutes. I got farther away and was drifting lower. The last time I saw it, it was probably a mile behind and a half mile below. I looked for it again in the periscope, over Kano, where he'd predicted it would be directly underneath and just a little bit south, but I couldn't see it then.

You wonder what your capabilities are gonna be. I had good control of the spacecraft and I reoriented to orbit attitude, put it back on ASCS and didn't drop into orientation mode or anything. I think you could have come off that booster with maybe running a controls check, that plus maybe another five minutes of just maneuvering to get the feel of this thing. I think you could have controlled accurately enough right at that stage to have gone back over and probably made an attempt to join up on something.

ATTITUDE CONTROL SYSTEM FUNCTION

The ASCS performed the turnaround maneuver properly and dropped into orbit mode. After Bermuda, I started into the controls check that we had practiced so many times. The test went just like clockwork. It was just like running it on the trainer again; the same rates and the same reaction of controls. I was really elated by the precision with which the controls check proceeded. It is quite an intricate little check until you practice it a number of times, and there are quite a number of switch and pull handle activations during this check. We had run through this so many times at the Cape, it was really a pleasure to see this thing work properly. The ASCS was working properly at that time. Following the controls check, I went back on ASCS and it held fine all through the first orbit.

Everything went according to plan during the first orbit. We had that long period on ASCS in order to hold an optimum attitude for radar and communications. Just as I got to the coast of Mexico, the capsule started drifting to the right in yaw and it would drift over to about 20 degrees, instead of the normal 30 degree limit, and then the high thruster would kick on and bat it back over to the left. It would overshoot to the left and then it would hunt and settle down again somewhere around zero. The spacecraft would then drift again to the right and do the same thing repeatedly. Sometimes it would do this just in yaw but at other times when it kicked into orientation mode in yaw, it would drop into orientation mode and all three axes would go through a period of fluctuation and the whole capsule was doing this until it gradually settled down. I let it do this three or four times and decided it was just too wasteful in fuel to let the thing continue, so every time the yaw would kick in, I would take over manually, re-center it slowly, and re-engage ASCS. It had that same characteristic every time I turned the ASCS back on for about some 15 to 20 minutes. On the second orbit, when I was pretty well over toward Africa, it reversed, and then it would drift to the left in yaw instead of the right. During the difficulty with yaw, when I would try to use low thrust to bring it back, there was no left low thrust. That's the reason the spacecraft would drift off course and hit orientation mode, where the high thrust had to take over. It seemed that later on if I held an attitude other than just zero-zero and minus 34 degrees for any period of time at all, any axis was liable to drift off in any direction and I never could pin this down to a pattern.

This is what consumed the bulk of my time for the last two orbits. I was trying to analyze the trouble with the ASCS, so we would know exactly what it was doing, but I was unable to pin this down to any exact pattern. The attitude indicators were most in error after I yawed around 180 degrees and held this attitude for several minutes, looking at these little bright spots at sunrise. I had had the capsule lined up in orbit attitude just before I made the 180-degree turn. I had caged and then uncaged the gyros after visually aligning the capsule to orbit attitude by reference to the horizon. Following my return to orbit attitude, the indicators were in error by 30 degrees in right roll, 35 degrees in right yaw, and 42 degrees pitch up, which puts it some degrees higher in pitch than it should have been at that point. I don't know whether this got reported or not. I felt that the gyros should have been aligned properly when I returned to orbit attitude unless there was some malfunction. During the majority of the first orbit, all the control systems functioned perfectly.

I could not pin down any pattern to the malfunction of the ASCS. I tried caging the gyros and getting everything lined up and then uncaging the gyros. Then I would try flying on the fly-by-wire to see if there was any connection between that system and the ASCS problem. It would still drift off in random fashion. So I would re-cage again, uncage, and try the same thing again on manual. There was just no pattern that could be established to it. Whatever the problem was, it was random.

I believe there was a period of learning to use the periscope and window during the flight. I think I wound up at the end of the flight much better able to discern yaw and set the capsule on the correct heading than I was at the beginning of the flight. Roll and pitch are so definite, that there is no doubt about controlling the attitude in these dimensions. You sit there and just drive the window where you want it. With accurate control on fly-by-wire it is very easy to position the capsule exactly where you want it. But this wasn't quite so true in yaw. Your reference in yaw is not as good. I found that the best way I could set up yaw was looking out the window and aligning with some object that was going straight away from me.

The procedure that seemed to aid this more than anything was pitching down. If I would pitch down to about 60 degrees or so where you have a pretty fair vertical view, then I could pick up clouds and land moving out from under me at a much more rapid rate than in orbit attitude – in which I was looking way off toward the horizon. When you are viewing the horizon the landmarks are much farther out and their apparent motion is considerably less if you are pitched down and are looking at them coming right out under your nose. When they are doing that, it's pretty easy to see what your yaw is. It's much more apparent that things are moving at an angle across the window when they are going at a greater speed. I wound up pitching down whenever I wanted to set yaw toward the end of the flight. The periscope I think just required a little more practice. At the beginning of the flight, it was a little bit difficult to pick up yaw rate, but I think by the end of the flight, I could probably set yaw on the periscope down to within a couple of degrees. I would say that this is about the tolerance of accuracy on the periscope, I don't think I could have got any closer on that. I think using the window you can get down to about the same accuracy maybe, possibly a little better. I think the big difference between the window and the periscope is probably the speed with which you can determine yaw. Looking out the window, you just come right around until it looks right, that's it, and it's very apparent to you because you are looking directly at the clouds and they are moving rapidly. On the periscope I seemed to have to do a lot more maneuvering and look and watch and track a cloud along the line and then make an adjustment, then track another cloud along the line and so on. It took longer to do it on the periscope than it did looking out the window.

You could determine yaw rate by star drift through the window. This took a little bit longer. If you were aligned pretty much on track it seemed a little bit more difficult to track the star and make your final accurate alignments in yaw. In fact, without watching it for a very, very long time it's apparent if you turned a little bit – yawed one way or the other as the drifting of the star field becomes more apparent. If you pick a particular star and track it, and it drifts rapidly while your rate indicator shows that the spacecraft is not yawing it is apparent that you are not quite on track. But, once you get this aligned pretty well on track, adjusting within the last 10 degrees of yaw is fairly difficult at night. The periscope gave no help at all in aligning with anything at night. You could see different patterns in the scope at night because we had a full Moon on those cloud decks but this wasn't much help – it's so dim through the periscope you couldn't really align with any particular spot and follow it as a yaw reference. So your yaw reference at night, if your instruments are not operating properly, is strictly out the window. You use the window in conjunction with the rate indicator when the yaw attitude is widely displaced from track. If you track a particular star, since you have no sensation of rate you must cross reference to the yaw rate indicator to be sure that the motion you see is not due to a capsule yaw rate. You have no sensation of turning unless you do look at the rate needle. Your rate indicators operate properly at all times. Aligning at night in yaw is not very easy. You can see cloud patterns out the window and this helps some. With the bright moonlight we had you could keep aligned by reference out the window on some particularly outstanding cloud formations. This was much more rapid than star reference. I think you could probably align out of the window, almost as rapidly as you could in the daytime, when you have a full Moon which illuminates the cloud formations.

ECS COOLING DURING WEIGHTLESSNESS

The comfort control, calibrations from the chamber runs at the Cape are not accurate at all, apparently. The setting of the suit circuit comfort control which provided cooling at the Cape during our ECS [Environmental Control System] runs was marked in red, right in the middle of the scales. By the end of the flight, I had increased this water flow on the suit circuit clear past the maximum setting on the scale which was 1.7 pounds per hour. My setting was beyond that and the steam exhaust temperature was still only down to about 47 degrees. I increased it slowly even beyond that point. I did not want to turn it way up for fear I would freeze the outlet up, so I kept edging it up slowly until I finally wound up with it clear above scale. On the other hand, the cabin temperature got up as high as 105 degrees at one time. I increased the flow slowly over a period of a full orbit and the cabin temperature gradually came down to about 95 degrees. When I tried to increase it even more to set the temperature down below 95 degrees, the excess water light came on, so I had to back it off. The water light went off in about ten minutes, so I turned up the water again. I turned the water off every time, as we were supposed to prevent hysteresis in the valve. When I would bring it back up again it would freeze. The excess water warning light was on and off five or six times. I never really achieved one setting which was close to the limit and still low enough to keep the excess water light off. I just didn't have a good feel for the cabin water position at all. It froze up rapidly on the position that was satisfactory in the chamber at the Cape. This, the planned setting on the suit circuit determined at the Cape, is too low, while on the cabin circuit the planned setting was too high.

SPACE FIREFLIES

Coming out of the night on the first orbit, at the first glint of sunlight on the capsule, I was looking inside the capsule to check some instruments for probably fifteen or twenty seconds. When I glanced back out the window, my initial reaction was that the capsule had tumbled and that I was looking off into a star field and was not able to see the horizon. I could see nothing but luminous specks about the size of the stars outside. I realized, however, they were not stars. I was still in the attitude that I had before. The specks were luminous particles that were all around the capsule. There was a large field of spots that were about the color of a very bright firefly, a light yellowish-green color. They appeared to vary in size from maybe just pinhead size up to possibly three-eighth of an inch. I would say that most of the particles were similar to first magnitude stars; they were pretty bright, very luminous. However, they varied in size so there would be varying magnitudes represented.

They were floating in space at approximately my speed. I appeared to be moving through them very slowly, at a speed of maybe three to five miles an hour. They did not center on the capsule as though the capsule was their origin. I thought first of the lost Air Force needles that are some place in space but they were not anything that looked like that at all. [In 1961 the USAF's Project West Ford launched 480 million copper dipole needles into orbit in an attempt to improve military communications by creating an artificial ionosphere above the Earth.] The other possibility that came to my mind immediately was that snow or little frozen water particles were being created from the peroxide water decomposition. I don't believe that's what it was, however, because the particles through which I was moving were evenly distributed and not more dense closer to the capsule.

As I looked out to the side of the capsule, the density of the field to the side of the capsule appeared to be about the same as directly behind the capsule. The distance between these particles would average, I would estimate, some eight to ten feet apart. Occasionally, one or two of them would come swirling up around the capsule and across the window, drifting very, very slowly, and then would gradually move off back in the direction I was looking. This was surprising, too, because it showed we probably did have a very small flow field set up around the capsule or they would not have changed their direction of motion as they did. No, I don't recall observing any vertical or lateral motion other than that of the particles that swirled around close to the spacecraft. It appeared to me that I was moving straight through a cloud of them at a very slow speed. I observed these luminous objects for approximately four minutes before the Sun came up to a position where it was sufficiently above the horizon that all the background area then was lighted and I no longer could see them.

After passing out of them, I described them as best I could on the tape recorder and reported them to the Cape. I had two more chances to observe them at each sunrise; it was exactly the same each time. At the first rays of the Sun above the horizon, the particles would appear. To get better observation of these particles and to make sure they were not emanating from the capsule, I turned the capsule around during the second sunrise. When I turned around towards the sunrise, I could see only ten percent as many particles as I could see when facing back toward the west. Still, I could see a few of them coming toward me. This proved rather conclusively, to me at least, that I was moving through a field of something and that these things were not emanating, at least not at that moment, from the capsule. To check whether this might be snowflakes from the condensation from the thrusters, I intentionally blipped the thrusters to see if I was making a pattern of these particles. I could observe steam coming out of the pitch down thruster in good shape and this didn't result in any observation of anything that looked like the particles. I had three good looks at them and they appeared identical each time. I think the density of the particles was identical on all three passes.

I would estimate that there were thousands of them. It was similar to looking out across a field on a very dark night and seeing thousands of fireflies. Unlike fireflies, however, they had a steady glow. Once in a while, one or two of them would come drifting up around the corner of the capsule and change course right in front of me. I think it was from a flow of some kind or perhaps the particles were ionized and were being attracted or repelled. It was not due to collisions because I saw some of them change course right in front of me without colliding with any other particles or the spacecraft. If any particles got in near enough to the capsule and got into the shade, they seemed to lose their luminous quality. And when occasionally, I would see one up very close, it looked white, like a little cottony piece of something, or like a snowflake. That's about the only description of them I have. There was no doubt about their being there because I observed them three different times for an extended period of time. I tried to get pictures of them, but it looks like there wasn't sufficient light emanating from them to register on the color film.

HIGH LAYER

I had no trouble seeing the horizon on the night side. Above the horizon some six to eight degrees, there was a layer that I would estimate to be roughly one and a half to two degrees wide. I first noticed it as I was watching stars going down. I noticed that as they came down close to the horizon, they became relatively dim for a few seconds, then brightened up again and then went out of sight below the horizon. As I looked more carefully, I could see a band, parallel to the horizon, that was a different color as moonlight on clouds at night. It was a tannish color or buff white in comparison to the clouds and not very bright. This band went clear across the horizon. I observed this layer on all three passes through the night side. The intensity was reasonably constant through the night. It was more visible when the Moon was up, but during that short period when the Moon was not up I could still see this layer very dimly. I wouldn't say for sure that you could actually observe the specific layer during that time, but you could see the dimming of the stars. But, when the Moon was up, you very definitely could see the layer, though it did not have sharp edges. It looked like a dim haze layer such as I have seen occasionally while flying. As stars would move into this layer, they would gradually dim; dim to a maximum near the center and gradually brighten up as they came out of it. So, there was a gradient as they moved through it; it was not a sharp discontinuity.

NIGHT SIDE OBSERVATIONS OF THE EARTH

Over Australia, they had the lights of Perth on and I could see them well. It was like flying at high altitude at night over a small town. The Perth area was spread out and was very visible and then there was a smaller area south of Perth that had a smaller group of lights [created by the citizens of Rockingham] but they were much brighter in intensity; very luminous. Inland, there were a series of about four or five towns that you could see in a row lined up pretty much east and west that were very visible. It was very clear; there was no cloud cover in that area at that time.

Knowing where Perth was, I traced a very slight demarcation between the land and the sea, but that's the only time I observed a coastline on the night side. Over the area around Woomera, there was nothing but clouds. I saw nothing but clouds at night from there clear up across the Pacific until we got up east of Hawaii. There was a solid cloud cover all the way.

In the bright moonlight, you could see vertical development at night. Most of the areas looked like big sheets of stratus clouds, but you could tell where there were areas of vertical development by the shadows or lighter and darker areas on the clouds.

Out in that area at night, fronts could not be defined. You can see frontal patterns on the day side. In the North Atlantic, you could see streams of clouds, pick out frontal areas pretty much like the pictures from earlier Mercury flights.

With the moonlight, you are able to pick up a good drift indication using the clouds. However, I don't think it's as accurate as the drift indications during the day. The drift indication is sufficient that you can at least tell what direction you're going at night within about ten or fifteen degrees. In the daylight over the same type clouds, you probably could pick up your drift down to maybe a couple of degrees.

The horizon was dark before the Moon would come up, which wasn't very long. But you can see the horizon silhouetted against the stars. It can be seen very clearly. After the Moon comes up, there is enough light shining on the clouds that the Earth is whiter than the dark background of space. Well, before the Moon comes up, looking down is just like looking into the 'black hole' of Calcutta.

There were a couple of large storms in the Indian Ocean. The Weather Bureau scientists were interested in whether lightning could be seen or not. This is no problem; you can see lightning zipping around in these storms all over the place. There was a great big storm north of track over the Indian Ocean; there was a smaller one just south of track and you could see lightning flashing in both of them; especially in the one in the north, it was very active. It was flashing around and you could see a cell going and another cell going and then horizontal lightning back and forth.

On that area, I got out the air glow filter and tried it. I could not see anything through it. This, however, may have been because I was not well enough dark adapted. This is a problem. If we're going to make observations like this, we're going to have to figure out some way to get better night adapted in advance of the time when we want to make observations. There just was not sufficient time. By the time I got well night adapted, we were coming back to daylight again.

DAYSIDE OBSERVATIONS

Clouds can be seen very clearly on the daylight side. You can see the different types, vertical developments, stratus clouds, little puffy cumulus clouds, and alto-cumulus clouds. There is no problem identifying cloud types. You're quite a distance away from them, so you're probably not doing it as accurately as you could looking up from the ground, but you can certainly identify the different types and see the weather patterns.

The cloud area covered most of the area up across Mexico with high cirrus almost to New Orleans. I could see New Orleans; Charleston and Savannah were also visible.

You can see cities the size of Savannah and Charleston very clearly. I think the best view I had of any area during the flight was the clear desert region around El Paso on the second pass. There were clouds north of Charleston and Savannah, so I could not see the Norfolk area and on farther north. I didn't see the Dallas area that we had planned to observe because it was covered by clouds but at El Paso, I could see the colors of the desert and the irrigated areas north of El Paso. You can see the pattern of the irrigated areas much better than I had thought we would be able to. I don't think that I could see the smallest irrigated areas; it's probably the ones that are blocked in by the larger sized irrigation districts which I saw. You can see the very definite square pattern in those irrigated areas, both around El Paso and at El Centro which I observed after retrofire.

The western part of Africa was clear. That is, the desert region where I mainly saw dust storms. By the time we got to the region where I might have been able to see cities in Africa, the land was covered by clouds. I was surprised at what a large percentage of the track was covered by clouds on this particular day. There was very little land area which could be observed on the daylight side. The eastern part of the United States and occasional glimpses of land up across Mexico and the desert area in Western Africa was all that could be seen.

I saw what I assume was the Gulf Stream. The water can be seen to have different colors. Another thing that I observed was the wake of a ship as I came over Recovery Area G at the beginning of the third orbit. I had pitched down to below retro-attitude. I was not really thinking about looking for a ship. I was looking down at the water and I saw a little vee. I quickly broke out the chart and checked my position. I was right at Area G, the time checked out perfectly for that area. So, I think I probably saw the wake from a recovery ship. When I looked back out and tried to locate it again, the little vee had gone under a cloud and I didn't see it again. The little vee was heading west at that time. It would be interesting to see if the carrier in Area G was fired up and heading west at that time.

I would have liked to put the glasses on and see what I could have picked out on the ground. Without the glasses, I think you identify the smaller objects by their surroundings. For instance, you see the outline of a valley where there are farms and the pattern of the valley and its rivers and perhaps a town. You can see something that crosses a river and you just assume that it's a bridge. As far as being able to look down and see it, and say that is a bridge, I think you are only assuming that it's a bridge more than really observing it. Ground colors show up just like they do from a high altitude airplane; there's no difference. A lot of things you can identify just as from a high flying airplane. You see by color variations the deep green woods and the lighter green fields and the cloud areas. I could see Cape Canaveral clearly and I took a picture which shows the whole Florida Peninsula; you [can] see across the interior of the Gulf.

Looking back at the Earth, colors and light intensities were the same as flying at high altitude in an airplane. You look down at the ground and you see the same colors and the ground looks the same as it does flying at 40,000 to 50,000 feet in an airplane. Now if you look off to the horizon, the view is completely different, of course. It is also different if you look up at the sky. But looking back at the ground it's just like you were in a high altitude airplane of some kind. I could see patterns on the ground and rivers. I particularly recall the area around El Paso, as one of which we had made a photographic study. The green irrigated area that goes up and down the valley each side of El Paso can be easily seen. It contrasted well with the desert area on each side of it. You could see squares of the irrigated land. I don't imagine that I was seeing the individual little irrigated plots that are only a few hundred yards long. I don't know what these were, maybe they were quarter sections. I could see rivers and lakes. You can see several cities, such as Savannah and Charleston, very clearly. Knowing that there is a river winding through an area, you interpret little blobs that you see as being bridges across the river. I think you can probably pick out areas down to maybe 100 yards long. We were figuring before that just using a gunsight on a fighter plane, you could pick out areas down to one mil. I think we figured that 1 mil at this altitude would be on the order of a hundred to one hundred and fifty yards. On the night side the visibility is pretty much dependent on whether the Moon is up or not. In one area off the east coast of Africa, before the Moon was up, I could see nothing in the way of cloud formations or anything at all. Reference then was strictly by the stars, but even without the Moon being up it didn't appear to be any problem at all seeing the horizon because the stars were so clear. Maybe you do have some very low intensity light that comes back off haze in the atmosphere that aids in seeing the night horizon. When there is no Moon out, looking right straight down is like looking into a black hole.

We had wondered about whether you could see lightning flashes at night and they were very visible. There were two big storm areas, one big general storm area that was just north of track off the east coast of Africa and another smaller storm center that was south of track and a little bit east of the first one. The lightning flashes in both of these storm areas were very visible. You could see the lightning going from cloud top to cloud top – sort of sheet lightning effect. You could see lightning in the clouds – lightning of the type you normally associated with clouds of vertical development where there is one single big flash rather than going from cloud to cloud. These were very visible. I tried the air glow filter in looking at that area but it was so soon after sunset that I don't think I was night adapted enough to really pick up air glow. The Indian Ocean [communications] ship said they set their flares off, but I did not see them.

SUNSET AND SUNRISE HORIZON OBSERVATIONS

At sunset, the flattening of the Sun was not as pronounced as I thought it might be. The Sun was perfectly round as it approached the horizon. It retained its symmetry all the way down until just the last sliver of Sun was visible. The horizon on each side of the Sun is extremely bright and when the Sun got down to where it was just the same level as the bright horizon, it apparently spread out perhaps as much as ten degrees each side of the area you were looking. Perhaps it was just that there was already a bright area there and the roundness that had been sticking up above it came down to where finally that last little sliver just matched the bright horizon area and probably added some to it.

I did not see the sunrise direct; only through the periscope. You cannot see that much through the scope. The Sun comes up so small in the scope that all you see is the first shaft of light. The band of light at the horizon looks the same at sunrise as at sunset.

The white line of the horizon is extremely bright as the Sun sets, of course. The color is very much like the arc lights they use around the pad. As the Sun goes on down a little bit more, the bottom layer becomes a bright orange and it fades into red; then on into the darker colors and finally off into blues and black as you get further up towards space. One thing that was very surprising to me, though, was how far out on the horizon each side of that area the light extends. The lighted area must go out some sixty degrees. I think this is confirmed by the pictures I took. I think you can probably see a little more of this sunset band with the eye than with a camera. I was surprised when I looked at the pictures to see how narrow looking it is. I think you probably can pick up a little broader band of light with the eye than you do with the camera. Maybe we need more sensitive color film.

CABIN LIGHTING

The sunlight coming in the window is very brilliant and intense, extremely bright, very clear brilliant white. It reminds me of the pad with the arc lights on. One of them shining in the window is a brilliant white light – that's just what it is like. Just like someone had an arc light right outside with the brilliant white light they put out. I didn't change the capsule lights for this at all. We had thought before the flight that we would probably not be able to look directly at the Sun. I sort of peeked around the edges of the window to try looking directly at the Sun without my eyes being protected. Previously, I had used the filter that we had for this purpose in the open position. By squinting the eyes it was possible to look directly at the Sun with no ill after effects at all. This would be very similar to looking directly at the Sun from the Earth. You don't accomplish anything by doing it since you can't see anything anyway with your eyes squinted down that far. But it is interesting at least that the Sun is not quite as intense in that regard as I thought it would be. It's not completely blinding if you happen to open your eyes to the Sun. I know a number of times when I'd be maneuvering the capsule, the Sun would come across the window and by squinting my eyes and looking to another part of the capsule, I could still see the instruments and see what I was doing, even though the Sun was coming directly in. It's very bright, I don't want to minimize it, but it does not stop all other activity when it happens to shine close to your face like we thought it might.

As far as the lighting of the capsule goes, the fingertip lights were very useful. We had thought from some of our practice in the procedures trainer the little finger tip lights we had might be very helpful and they were. I used them numerous times to look at the charts. They were also useful before my night adaptation was complete, for making gauge readings or trying to read out a report. The instrument panel is not evenly lighted in the capsule. You really need edge lighting on some of the gauges to make them readable. We have flat red lighting on gauges that are rather far away from the light sources, so that the illumination is dim.

Another factor in capsule lighting is that you need to be dark adapted to where you can really make the maximum observation of the sky shortly after sunset, or determine yaw position from a cloud formation out the window immediately after sunset. To do this, you would have to adapt your eyes a lot more in advance of sunset than we had planned to do on this flight. I tried for maybe a minute or so before sunset. We had an eye patch and I tried it on the first orbit. It was sort of a jury rig item that we put together at the last minute with tape around the edges to hold it in place. The tape just didn't stick the way it should have. It kept coming loose and I kept changing its position. I thought from our previous practice that it would work all right, but it wasn't worth much. I gave up on it and stuck it over on the emergency O_2 rate box. It wasn't much good. So I tried then just to shut my eye and adapt one eye while we were coming up to the dark side so that I'd have pretty good reference after we went through sunset. I just wasn't trying it far enough ahead because to make any real observations or really pick up yaw reference rapidly, you have to be night adapted farther ahead than I was.

RETROFIRE

Retrorockets were fired right on schedule just off California and it was surprising coming out of the zero-g field that the retrorockets firing felt as though I were accelerating in the other direction back toward Hawaii. However, after retrofire was completed, when I could glance out the window again, it was easy to tell, of course, which way I was going, even though my sensations during retrofire had been that I was going in the other direction. I made retrofire on automatic control. Apparently, the solid-on period for slaving just prior to retrofire brought the gyros back up to orbit attitude, because they corrected very nicely during that period. The spacecraft was just about in orbit attitude as I could see it from the window and through the periscope just prior to retrofire. So, I feel that we were right in attitude. I left it on ASCS and backed up manually and worked right along with the ASCS during retrofire. I think the retro-attitude held almost exactly on and I would guess that we were never more than 3 degrees off in any axis at any time during retrofire.

I went to manual control after retrofire because I set my roll rate in at about the time we had reached a quarter or a half g. The early part of the damping was okay. I didn't feel that I had all the authority that I wanted when the oscillations started building up and I went to fly-by-wire in addition to the manual; I don't think this was until after 0.05-g. I'd like to reserve comment on that until we look back to the record because I'm not real positive where I went to fly-by-wire in addition to manual.

RE-ENTRY

Following retrofire, a decision was made to have me re-enter with the retro-package still on because of the uncertainty as to whether the landing bag had been extended. I don't know all the reasons yet for that particular decision, but I assume it had been pretty well thought out and it obviously was. I punched up 0.05-g manually at a little after the time it was given to me. I was actually in a small g-field at the time I pushed up 0.05-g and it went green and I began to get noise, or what sounded like small things brushing against the capsule. I began to get this very shortly after 0.05-g and this noise kept increasing. Well before we got into the real heavy fireball area, one strap swung around and hung down over the window. There was some smoke. I don't know whether the bolt fired at the center of the pack or what happened. The capsule kept on its course. I didn't get too far off of re-entry attitude. I went to manual control for re-entry attitude through the high-g area.

Communications blackout started a little bit before the fireball. The fireball was very intense. I left the shutters open the whole time and observed it, and it got to be a very, very bright orange color. There were large flaming pieces of what I assume was the retro-package breaking off and going back behind the capsule. This was of some concern, because I wasn't sure of what it was. I had visions of them possibly being chunks of heat shield breaking off, but it turned out it was not that.

There was no doubt when the heat pulse occurred during re-entry. I had the window shutters open and could see the glow outside. The ECS cooling was okay during the early parts of re-entry and I didn't feel particularly hot during the heat pulse. However, about the time we were getting down to around 75,000 or 80,000 feet it got very noticeably warm. It went from the same comfortable level I had been used to up to where it was uncomfortably warm in about fifteen to twenty seconds. I never had been able to cool the capsule down as much as I wanted, so I was not precooled when starting into the re-entry. The capsule stayed uncomfortably warm all the way down through drogue and main chute. When I was on the water I was sweating profusely.

The g's were no problem. This is similar to centrifuge runs as far as the g tolerance goes. I could still talk at 8 g's and although you apparently didn't hear me, I was communicating through that area of g buildup through the communications blackout period. We had all been able to communicate on the centrifuge to varying levels up to 14 g's, although at 14 they were pretty much grunts and some largely unintelligible things. But there was limited communication up to about 14 g's so communicating at 8 g's was no particular problem. I think our experience on the centrifuge was very similar to this as far as acceleration goes.

There were oscillations through the main g pulse and these were easily damped out. Following the main g pulse there were just minor oscillations. When we got down to about 50,000 or 55,000 feet the oscillations were divergent and they built up to an estimated plus and minus seventy to eighty degrees. The oscillations after peak-g were more than I could control with the manual system. I was damping okay, but it just plain overpowered me and I could not do any more about it. I switched to Aux Damp as soon as I could raise my arm up after the g pulse to help damp and this did help some. However, even on Aux Damp, the capsule was swinging back and forth very rapidly and the oscillations were divergent as we descended to about 35,000 feet. The frequency of oscillation was about one cycle per second. At this point, I elected to try to put the drogue out manually, even though it was high, because I was afraid we were going to get over to such an attitude that the capsule might actually be going small end down during part of the flight, if the oscillations kept going the way they were, so that if we fired the drogue it might get wrapped up in the capsule. And just as I was reaching up to pull out the drogue on manual, it came out by itself. The drogue did straighten the capsule out in good shape. I believe the altitude was somewhere between 30,000 and 35,000 at that point. I came on down; the snorkels, I believe, came out at about 16,000 or 17,000. The periscope came out. There was so much smoke and dirt on the windshield that it was somewhat difficult to see. Every time I came around to the Sun – for I had established my roll rate on manual – it was virtually impossible to see anything out through the window. The linear g impulse during drogue and main is not a big acceleration that bothers you. The opening of the chute was a pretty good jolt but was a little mushier or softer that I thought it would probably be due to the stretchy nylon risers.

The capsule was very stable when the antenna section jettisoned. I could see the whole recovery system just lined up in one big line as it came out. It unreeled and blossomed normally; all the panels and risers looked good. I was going through my landing checklist when the Capsule Communicator called to remind me to deploy the landing bag. I flipped the switch to auto immediately and the green light came on and I felt the bag release. I was able to watch the water coming towards me in the periscope. I was able to estimate very closely when I would hit the water.

Landing was a little bit sharper than I had figured it would be – the impact bag was a heavier shock than I had expected – but it did not bother me. I don't know whether you had this type of sensation or not. I had thought that this nice big air ride bag back there as being a reasonably soft landing, but this was a pretty big jolt. It wasn't anything that is debilitating or anything like that, but it was just a little more than I had anticipated – a pretty good kick in the back, but nothing anywhere near severe enough to cause any difficulty. The biggest difference is that you don't have any tumbling on starting or stopping like you do on the centrifuge where you have been going in rotary motion and when you stop or accelerate rapidly you have the feeling of going head-over-heels or heels-over-head, depending on which way you're going. There's none of this during re-entry.

LANDING

Yes, I had gotten into position so that in case I was not estimating altitude properly through the scope and it sneaked up on me, I would be ready for it – so I was back and pretty well braced. The closer I got to the water, the more detail I could pick out through the scope. I used the little hand mirror a couple of times but there was so much junk on the window that I couldn't get nearly as good a view of the water with the mirror as I did through the scope. I finally gave up trying to estimate my height with the mirror. One reason that I wanted the mirror was that I thought it would give us our best estimation of height above the water, but I just couldn't see enough through the window to make a good estimation. The closer I got to the water, though, the better the indications were through the periscope, and, just before impact, I could see the little whitecaps and every little wave detail and you can get braced very adequately.

I think the capsule was pretty well lined up straight below the chute. It was a straight g pulse into the couch but we had a pretty fair oscillation set up after that. After main chute, the oscillation wasn't large but we were swinging back and forth for plus and minus twenty degrees at the maximum, and the closer we got to the water the more this damped out. This is similar to some of the pictures we've seen of capsules coming down where they had more oscillation shortly after opening than it did close to the water. The chute was very stable by the time I got down to the water. I think we hit with almost no apparent swinging motion. At impact I could see no drift in looking out the periscope – I was watching through the scope and I could tell almost to the second when I was going to hit, because I could see the waves coming up very close right at the last. I could see every little whitecap and the detail on them, and just before landing I braced myself. So you can get a pretty good indication of height just by looking out the periscope. The action of the capsule at impact; it rolled to my right and down a little bit toward my feet down toward the hand controller or the number one battery. It went to that side momentarily and then collapsed over on its side like, and then there was water all over everything. The capsule popped back up and assumed various angles from there on.

When I had bobbed back up again, the rescue aid light was red. I operated it manually and I didn't feel the main chute disconnect. I wasn't positive that the chute had disconnected for about ten or fifteen seconds, then I saw it through the scope in the water and I knew it was disconnected at that time – or at least it wasn't pulling the capsule any more. I didn't have any sensation of the reserve chute blowing out at all, or of firing, and I assume that it did. This package was lying in the water alongside the capsule – I could see that through the periscope – but I don't recall hearing any big firing of the reserve chute and it's surprising because these other pyros operating earlier were so audible that I would think I would have heard that one in particular. I operated this switch pretty rapidly after impact, though I think I caught it bobbing back up before we'd been on the water very long.

There was quite a bit of noise from the water hitting the side and it may have blanketed some of this other sound from the reserve chute being popped out.

No, there was nothing flying around inside at all during impact. The g loading at impact – that's a difficult one to estimate because it's mainly a shock g and those are extremely hard to estimate. I would guess the initial shock is probably 12 to 14 g but it's very short; probably not over one quarter or one half second duration.

There was no leakage at all. I looked everywhere I could look to see if there was any leakage, especially after these chunks had been coming off during re-entry. I had visions of these being possibly either a broken heat shield or something there that might puncture the capsule or burn through it a little bit. Actually, back in that area the only place you can look is in the area surrounding you in the upper part of the capsule where a leak probably would not occur.

RECOVERY

My initial physical contact between the capsule and the ship was a very gentle rubbing up against the side of the ship; I could tell when it happened, of course, but it was very gentle and [speaking on UHF with the ship's captain] he immediately said they had the shepherd's crook on it. There was very little delay in starting their hoisting operation. They pulled the capsule up part way out of the water and then let water drain out of the bag in that position. They pulled it up rather slowly for a little while, then hoisted it right on up and stopped in a position out of the water. I thought they did a fine, expeditious job of picking it up. Someone told me he had cut the antenna. He was keeping me very well informed all during his approach. During the capsule pickup I got one real good bump, it was probably the most solid bump of the whole trip. I understand that a line slipped or they didn't have one of their stabilization lines pulled up. The ship rolled and swung the capsule out away from the ship and then rolled back and swung the capsule against the ship. It jammed me over to the side of the couch. But it wasn't anything that would hurt me.

After we were on deck, there was some delay while they were doing something up on top – I never did find out for sure what they were doing. They asked if I could hear the men talking down from up on top. I could not; there was quite a bit of inverter and other noises still going on in the capsule. I was still in good communications with the ship, however, on UHF. I took the panel loose – I kept the suit hose connected and worked over and couldn't quite reach the last restraining strap for that hose extension that was stowed; remember it is way down by the end of the wire bundle on the right. I worked a little bit trying to get down to the extension hose and still keep the vent hose connected. I didn't want to disconnect the vent hose until I had to. I probably added more heat to the system in that period of straining against the vent hose restraining strap than if I'd gone ahead and disconnected the hose and reached down and picked up the extension. But I finally pulled the extension loose without disconnecting the vent. I was working my way up to sitting on the right side of the capsule, had my head up towards the back of the [top] hatch. I then disconnected the suit hose and was in the process of hooking up the hose extension when I decided to egress by the side hatch. I was really hot by this time. I was pouring sweat, and I decided the best thing at that point to keep me in a little better shape was to come on out rather than come up through the top. If I had had to get out through the top I could have, but I was extremely hot at that time and with the means of getting out I didn't see any point in fighting that one any further. I don't know whether you heard our communications with the ship or not at that time. I think he asked me whether I wanted to blow it or not. I said I did and to let me know when everybody was clear outside. He called me back and told me when everybody was clear.

I took the cap off and the pin out and positioned my hand with my knuckles beside the plunger, turned my head back away and shielded my head with my arm and then bumped the plunger with my hand. That was a mistake because the plunger kicked back out and cut both of my knuckles through the gloves and threw my arm back over in the capsule. My knuckles were bleeding down both fingers when I finally undressed later on. I was afraid, at first, I might have broken them. I highly advise if anyone ever has to use that plunger again they do it with the meaty part of their hand or possibly take the swizzle stick and rap it from the side. The hatch made a good loud bang when it left but it wasn't uncomfortably loud. I still had the helmet on at the time so that undoubtedly attenuated a lot of the noise. I noticed after the hatch blew that quite a bit of grey paint from the inside of the hatch area had flaked off and settled inside the capsule. Well, this is very much like climbing out of the procedures trainer or climbing out of the capsule on the pad. There were two men back behind, helping lift me out and keeping me up above the edge of the hatch so that we didn't rip the suit. Egress was completely normal; I pulled my legs around in the same position we have for getting in and out. I asked one of the men to hold his arm up so that I could hang onto it. We have found climbing in and out of the capsule is a lot easier if you have a handle up above the hatch or someone's arm to hold on to so you can hoist yourself and slide right in and out, and he did that.

SUMMARY

In summary, my condition is excellent. I am in good shape; no problems at all. The ASCS problems were the biggest I encountered on the flight. Weightlessness was no problem. I think the fact that I could take over and show that a pilot can control the capsule manually, using the different control modes, satisfied me most. The greatest dissatisfaction I think I feel was the fact that I did not get to accomplish all the other things that I wanted to do. The ASCS problem overrode everything else.

Appendix 4

Naming that tune

In October 1957, three months after he had broken the trans-America speed record flying a Crusader, Maj. John Glenn received a surprise invitation to appear on the Tuesday evening CBS game show, *Name That Tune*. Hosted by the affable George DeWitt, contestants had to quickly identify songs being played by a studio orchestra, ring a bell or press a buzzer before the opposing team, and correctly name the tune. Wearing his dress uniform with a chestful of medals, Glenn had been paired off with a sprightly freckle-faced lad named Eddie Hodges. "They always teamed people up on that program," Glenn recalled during an interview, "and Eddie Hodges, later the juvenile lead in the original cast of *Music Man*, was my partner on that. He was a ten-year-old boy." They made a good and popular team. Over the next five weeks they would participate in what was called the Golden Medley Marathon, which was worth $25,000 to the winning team, naming such tunes as *The Chicken Reel* and *Far Away Places*.

Eddie Hodges' family had relocated from Mississippi to New York in 1952, when he was five years old. One day a female talent scout saw him strolling through Central Park, which led to his appearance on the top-rated quiz show.

Eclipsing all other news on 4 October was the unexpected revelation that the Soviet Union had put *Sputnik* into orbit around the Earth, as the world's first artificial satellite. During the show that evening, DeWitt asked Glenn what he thought of the Russian satellite.

"Well, to say the least, George, they're out of this world," he responded, which garnered a laugh from the studio audience. Then, in a more serious vein, Glenn provided an inkling of the diplomatic introspection for which he would soon become famous, and foretold a future in which he would fully participate. "This is really quite an advancement for not only the Russians but for international science; I think we all agree on that. It's the first time anybody has ever been able to get anything out that far in space and keep it there for any length of time, and this is probably the first step toward space travel or Moon travel; something we'll probably run into maybe in Eddie's lifetime here at least."

DeWitt then asked Eddie Hodges if he would like to take a trip to the Moon.

"No, sir," came the quick response, accompanied by a bright smile. "I like it fine right here."

© Springer International Publishing Switzerland 2015
C. Burgess, *Friendship 7*, Springer Praxis Books, DOI 10.1007/978-3-319-15654-5

Eddie Hodges and U.S. Marine pilot John Glenn on *Name That Tune* in 1957. (Photo: John Glenn Archives, The Ohio State University)

The duo went on to appear in several broadcasts of the show. As Glenn later wrote in his 1999 memoir: "Eddie and I made it through the first two rounds and won $10,000, pushing a button ahead of the clock when we recognized the tune. We won the next week, and the next, adding $5,000 a week to our winnings. .... If Eddie and I made it through the fifth and final round of the Golden Medley Marathon, we would split the grand prize of $25,000. Even half of $25,000 was an astounding amount of money to a Marine pilot."[1]

They finally won the grand prize of $25,000, which they shared. "After taxes, I wound up with $8,000," Glenn later recalled.[2] He put his take into a college fund for his children and a home electric organ for Annie.

John Glenn hugs Eddie Hodges after they won $25,000, while host George DeWitt congratu-
lates the popular duo. (Photo: John Glenn Archives, Ohio State University)

By chance, a big fan of *Name That Tune* was the wife of Meredith Wilson, who was
then casting for the Broadway musical, *The Music Man* and was looking for a juvenile to
play the singing role of a young boy named Winthrop Paroo. "We found out later that
when she saw me on *Name that Tune* she called her husband and told him she had found
his Winthrop," says Hodges.[3] He auditioned, got the role, and spent the next year making
405 appearances in the hit musical. "Annie and I were there on opening night to hear Eddie
perform," Glenn later wrote in his memoir, "including the wonderful solo 'Gary, Indiana.'"[4]

In 1959, aged 12, Hodges became Mississippi's first Grammy Award winner for being
part of the original Broadway cast soundtrack album of *The Music Man*. That same year,
in which John Glenn was selected as a Mercury astronaut, Hodges made his film debut in
Frank Capra's *A Hole in the Head* along with Frank Sinatra, in which he and Sinatra performed a song called *High Hopes*. He also starred in the 1960 Michael Curtiz film, *The
Adventures of Huckleberry Finn* and in 1960 even had his own television musical show,
The Secret World of Eddie Hodges.

In 1961, at the age of just 14, Hodges had a huge hit record with his version of the Isley Brothers' song *I'm Gonna Knock On Your Door*. The following year he also scored a lesser hit with *(Girls, Girls, Girls) Made to Love*, originally recorded by The Everly Brothers. He then recorded songs for other record labels, while keeping his film career going by starring with Hayley Mills in *Summer Magic* (1963) and another Disney film, 1967's *The Happiest Millionaire*. From then on Hodges enjoyed a continuous stream of work in movies, television and recording. Guest appearances on network TV programs included *Bonanza, Gunsmoke, Cimarron Strip*, and *The Dick Van Dyke Show*.

Hodges went on to work as a union musician, record producer, song writer, and music publisher. In collaboration with Tandyn Almer he wrote and published several songs and owned his own music publishing business, but he decided to retire from acting in the early 1970s to become a mental health counselor back in his native Mississippi. He continues to write songs today but is no longer involved in the music industry and once again resides in Mississippi.

John Glenn and Eddie Hodges can both lay claim to having enjoyed truly stellar careers since their unforgettable collaboration on *Name that Tune*, which was revisited in the 1983 movie, *The Right Stuff*, starring Ed Harris and child actor Erik Bergmann, in their respective roles as the astronaut and the talented schoolboy.

As an interesting slant on the film, Erik Bergmann was asked to reflect on what he describes as "one of my favorite childhood memories" and how it came about.

"I was cast as Eddie Hodges when I was ten years old, mid-winter of 1981. I was living in a San Francisco suburb and had started auditioning professionally in the city a year or two before, after taking an on-camera acting class. My memory is that I had one initial audition and two callbacks. The final callback was with Phil Kaufman himself at Francis Ford Coppola's Zoetrope building, an odd, green, triangular structure in San Francisco's North Beach neighborhood (similar to the Flatiron building in NYC).

"What I remember most clearly from this audition was singing for Kaufman, and my mother later observed (as she actually remembered watching Eddie occasionally belt out songs on *Name That Tune* in her childhood) that this may have been what clinched the job for me. The song I performed for Kauffman was one I had written a year or so before about a street performer, entitled *The Music Man*. It wasn't until years later that I discovered Eddie had originated the role (later played by Ron Howard in the movie) of Winthrop Paroo in the Broadway production of Meredith Wilson's *The Music Man*. I was completely unaware at that time of the musical's existence, let alone its connection to Hodges, but since learning of it, I've often wondered if this synchronous little tidbit didn't play into my casting as well.

"Regardless, after getting word I'd booked the job, I remember going to a large loft space filled with costumes and props laying on large wide tables and hanging from walls, where I was fitted for the Boy Scout uniform I wore in the film – vintage, right down to the shoes which were never seen on camera.

Erik Bergmann as Eddie Hodges, and Ed Harris as John Glenn. (Still images from movie *The Right Stuff*)

"The day before the shoot, my blond hair was dyed the appropriate Eddie Hodges red, even though the scene was to be in black and white. We shot the following morning on a large soundstage in the industrial part of San Francisco which I later learned was somewhat infamous among the local film community as a favored location for adult filmmakers. I remember seeing my name written in Sharpie on a strip of white gaffer's tape on the door of my broom cupboard-sized dressing room and how special it made me feel.

"I especially remember spending quite some time on set chatting with Ed Harris while they got lighting and shots squared away. I also remember people marveling over and over about how much he looked like the real John Glenn, though I didn't understand how accurate that was as I had no real awareness of Glenn at the time."[5]

Hollywood producer Robert Chartoff worked on the movie *The Right Stuff*, and he remembers Ed Harris from the first moment he turned up to audition for the role in front of him and writer-director Philip Kaufman in 1979. "Ed Harris walked into the office, and we looked at him and couldn't believe that such a person existed. He was not only a wonderful actor but looked so much like John Glenn. And of course we looked at each other and said, 'Oh my God, this is the guy we want.' I said to Phil, 'Please, don't let this guy get hit by a car. At least, not until after the picture is made.'"[6]

Once selected, Harris threw himself into the role with zeal. Some of the dramatic re-entry scenes took part on a stage with a four-foot-tall model of *Friendship 7*. "I knew that capsule inside and out," he recalled. "I knew what all the gauges were and everything. You're just using your imagination. Like a kid, you know, climbing under a bunch of blankets pretending you're going to the Moon."[7]

Erik Bergmann had a lot of admiration for Harris. "Ed was very kind to me, speaking as with an equal. I recall his sharing that he had an upcoming guest part on the popular cop-drama, *CHiPs*, a show I already watched religiously, and which I watched with even more excitement when, a couple of weeks later, Harris' episode aired. I remember calling out to my parents "Hey, it's ED!" when he came on screen (as a bad guy, if memory serves), as if he was just another buddy. And this stands out in my memory because at that point he really was. *The Right Stuff* was his springboard into Hollywood stardom, but when I knew him, he was just a cool older (to me) actor who had treated my ten-year old self like an adult.

This slightly blurred Polaroid photo of the two actors is one of Erik Bergmann's most prized childhood possessions. (Photo courtesy of Erik Bergmann)

"The shoot only took a few hours. We were released by noon, but were invited to join the crew for lunch in a nearby parking lot where catering had been set up. This memory stands out for me as well, because after stopping by and enjoying a little food and conversation, and just before getting ready to head home, I was approached by a tall dark haired man who told me he was also in the film, and that he and another actor would be the people watching my scene on TV in the movie. As my mother and I walked away she asked me if

I recognized the guy I'd been talking to, and I admitted at first that I didn't. 'That's Jeff Goldblum. You know, the guy you watched on *Ten Speed and Brown Shoe*?' This was also a favorite show (at that point no longer on the air), and once I realized who he was, my mom encouraged me to go ask for his autograph. Shyly I approached Goldblum, and upon hearing my request he immediately responded, 'Why, of course! But only if I can have yours!' So we exchanged autographs and I left a very happy little boy. Somehow I doubt he still has mine, but I think I still have his tucked away in some scrap book or another. It's certainly one of my favorite memories of that day.

"As mine was one of the first scenes shot of the film, and post-production turned out to be a nightmare (or so I've heard), I had to wait almost three years for the release. I didn't even know if I was still in the film, as I could have easily ended up on the cutting room floor. But not long before the release, I was standing in a crowd with some school chums watching a film shooting on the street near our school (Gene Wilder directing/acting in *The Woman In Red*) and a crew member walking past me stopped and said 'Hey, I remember you! I worked on *The Right Stuff*. Your scene looks great!' Not only was I stunned to be recognized, but I later realized this probably meant I had made it into the film! Sure enough, not too long after I found myself in a packed audience at the city's largest film venue, the Northpoint Theater (now sadly defunct), face-to-face with my 30-foot-high alter ego.

"It's funny; I remember going into the theatre worried, knowing about the film's three-plus-hours running time and knowing my scene was somewhere in the middle, and also knowing – let's be honest – that I was there mainly to see myself, and that I might be bored slogging through the rest of the movie. But I can honestly say I needn't have worried. I walked out thoroughly entertained from beginning to end, and, to this day, it is one of my favorite films, regardless of my role in it. Okay, I'm probably a *little* biased, but I honestly

Erik Bergmann today. (Photo courtesy of Erik Bergmann)

believe it's a great film and an important one, and to have played a small part in something that has turned out to be meaningful to so many people, is a feather I gratefully carry in my cap to this day."[8]

Following his role in The Right Stuff, *Erik Bergmann continued working in San Francisco and the Bay Area until he relocated to New York City where he currently lives and works as a full-time union actor.*

REFERENCES

1. Frank Van Riper, *Glenn: The Astronaut Who Would be President*, Empire Books, New York, NY, 1983
2. *Ibid*
3. Hattiesburg Eye Clinic online article, "Famed Child Star's Vision Restored," 30 August 2013. Website: *http://hattiesburgeyeclinic.blogspot.com.au/*
4. John Glenn with Nick Taylor, *John Glenn: A Memoir*, Bantam Books, New York, NY, 1999
5. Erik Bergmann email correspondence with Colin Burgess, 15 October-11 November 2014
6. Alex French and Howie Kahn, "Punch a Hole in the Sky: An Oral History of the Right Stuff," online at *http://www.wired.com/2014/11/oral-history-of-right-stuff/*
7. *Ibid*
8. Erik Bergmann email correspondence with Colin Burgess, 15 October – 11 November 2014

Appendix 5

The Mercury pressure suit

Excerpts from the "Life Support Systems and Biomedical Instrumentation" subsection of NASA Report Results of the First Manned United States Orbital Space Flight, *NASA Manned Spacecraft Center (now the Johnson Space Center), Houston, Texas. Subsection authors: Richard S. Johnston, Asst. Chief, Life Systems Division, NASA Manned Spacecraft Center; Frank H. Samonski, Jr., Maxwell W. Lippitt, and Matthew I. Radnofsky (all from the MSC Life Sciences Division).*

INTRODUCTION

The pressure suit used in the MA-6 flight was developed from the U.S. Navy MK-IV full pressure suit manufactured by the B. F. Goodrich Co. This basic suit was selected by NASA in July 1959 for use in Project Mercury after an extensive evaluation program of three full pressure suits. This initial suit evaluation was conducted by the U.S. Air Force Aerospace Medical Laboratory, Aeronautical Systems Division. Many design changes have been made to the suit since the start of the Mercury program and, indeed, changes and modifications are still being investigated to provide as good a suit as possible for the Project Mercury flights.

The full pressure suit consists of five basic components; the suit torso, helmet, gloves, boots, and undergarment.

© Springer International Publishing Switzerland 2015
C. Burgess, *Friendship 7*, Springer Praxis Books, DOI 10.1007/978-3-319-15654-5

Suit engineers at B. F. Goodrich demonstrate a most unusual test of the spacesuit's mobility. (Photos: Goodrich Corporation)

PRESSURE-SUIT TORSO

The suit torso … is a closely fitted coverall tailored for each of the astronauts. It covers all of the body except for the head and hands. The torso section is of two-ply construction: an inner gas-retention ply of neoprene and neoprene-coated nylon fabric, and an outer ply of heat-reflective, aluminized nylon fabric. The helmet is attached to the torso section by a rigid neck ring. A tie-down strap is provided on this neck ring to prevent the helmet from rising when the suit is pressurized. Straps are also provided on the torso section for minor sizing adjustments of leg and arm length and circumferences and to prevent the suit from ballooning when pressurized.

Donning and doffing of the suit is provided through a pressure-sealing entrance zipper which extends diagonally across the front of the torso from the left shoulder down to the waist. Two frontal neck zippers and a circumferential waist zipper are also provided for ease in donning and doffing.

The pressure suit ventilation system is an integral part of the torso section. A ventilation inlet port is located at a point just above the waist on the left side of the torso section. This inlet port is connected to a manifold inside the suit where vent tubes lead to the body extremities. These tubes are constructed of a helical spring covered by a neoprene-coated nylon fabric that contains perforations at regular intervals. Body ventilation is provided by forcing oxygen from the environmental control system into the inlet and distributing this gas evenly over the body. The ventilation system in the Mercury pressure suit was especially developed to insure compatibility with the environmental control system.

Goodrich engineers begin the manufacture of John Glenn's personal spacesuit. Note the humorous inscription on the torso panel. (Photo: Goodrich Corporation)

The suit torso section contains several items which have been developed specifically for Project Mercury. They are as follows:

Bioconnector – The bioconnector provides a method for bringing medical data leads through the pressure suit. The bioconnector consists of a multipin electrical plug to which the biosensors are permanently attached, a receptacle plate mounted to the suit torso section, and an outside plug which is connected to the spacecraft instrumentation system. With this system, the biosensor harness is fabricated with the bioconnector as an assembly and no additional electrical connectors are introduced into the transducer system. In operation … the male internal plug is inserted inside the suit receptacle and locked into place. The internal plug protrudes through the suit to allow the spacecraft plug to be attached. The bioconnector system has proven to be a much more satisfactory connector than the previously used biopatch.

Neck dam – A conical rubber neck dam is attached to the torso neck ring. …. The purpose of this neck dam is to prevent water from entering the suit in event of water egress with the helmet off. The neck dam is rolled and stowed on the outside of the neck ring disconnect. After the astronaut removes the helmet in preparation for egress, he unrolls the neck dam until it provides a seal around his neck.

Pressure Indicator – A wrist-mounted pressure indicator is worn on the left arm. This indicator provides the astronaut a cross-check on his suit-pressure level. The indicator is calibrated from 3 to 6 psia [pounds per square inch absolute].

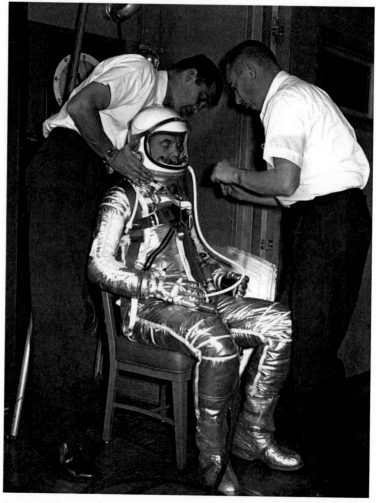

Making final adjustments to Glenn's spacesuit. (Photo: Goodrich Corporation)

Blood-Pressure Connector – A special fitting is provided on the suit torso which permits pressurization gas to be fed into the blood-pressure cuff. A hose leading from the cuff is attached to this connector during suit donning. After astronaut ingress into the spacecraft the pressurization source is attached to the connector on the outside of the suit.

HELMET

The helmet assembly ... consists of a resinous, impregnated Fiberglas hard shell, an individually molded crushable impact liner, a ventilation exhaust outlet, a visor sealing system, and a communications system.

The helmet visor sealing system consists of a pivoted Plexiglas visor, a pneumatic visor seal, and an on-off visor valve. Closing the visor actuates the valve and causes automatic inflation of the visor seal. The visor seal remains inflated until a deflation button on the valve is manually actuated by the astronaut. The valve has provision for attachment of the visor-seal gas-supply-bottle hose.

The helmet communication system consists of two independently wired AIC-10 earphones with sound attenuation cups and two independently wired AIC-10, newly developed, dynamic, noise-cancelling microphones. The microphones are installed on tracks which allow them to be moved back from the center of the helmet to permit eating and proper placement.

GLOVES

The gloves attach to the suit torso at the lower forearm by means of a detent ball-bearing lock. The gloves have been specially developed for Project Mercury to provide the maximum in comfort and mobility. Early centrifuge programs dictated the requirements for this development. Poor mobility in wrist action when the suit is pressurized caused an impairment in the use of the three-axis hand controller.

A pressure-sealing wrist bearing was incorporated to improve mobility in the yaw-control axis. The one-way stretch material on the back of the gloves improves mobility in the pitch and roll axes.

The gloves have curved fingers so that when pressurized the gloves assume the contour of the hand controller. The glove, like the torso section, has a two-ply construction – the inner gas retention ply and an outer restraint ply. The inner ply is fabricated by dipping a mold of the astronaut's hand into Estane material. The outer ply is fabricated from one-way stretch nylon on the back of the hands and fingers and a neoprene material injected into a nylon fabric in the palm of the gloves to prevent slippage in turning knobs, and so forth. Lacings are provided on the back of the glove to allow for minor adjustments. Two wrist restraint straps are provided to form break points and thereby improve pressurized glove mobility.

Miniature needle-like red finger lights are provided on the index and middle fingers of both gloves. Electrical power is supplied to the miniature lights by a battery pack and switch on the back of the gloves. These lights provide instrument panel and chart illumination before the astronaut is adapted to night vision.

Glenn demonstrates the miniature fingertip lights built into his gloves. (Photo: NASA)

BOOTS

Lightweight, aluminized, nylon-fabric boots with tennis-shoe-type soles were specially designed for the Mercury pressure suit. These boots resulted in substantial weight savings, provided a comfortable boot for flight, and a flexible friction sole which aids in egress from the spacecraft.

UNDERGARMENT

The undergarment is a one-piece, lightweight, cotton garment with long sleeves and legs. Thumb loops are provided at the sleeve ends to prevent material from riding up the arms during suit donning. Ventilation spacer patches … of a trilock construction are provided on the outside of the undergarment to insure ventilation gas flow over certain critical areas of the body.

As NASA suit technician Joe Schmitt prepares Glenn's space suit, the astronaut is seen wearing the lightweight undergarment. (Photo: NASA)

PRESSURE-SUIT SUPPORT

Prior to and after astronaut donning of the pressure suit the complete assembly was pressurized and leak checked at 5 psig [pounds per square inch gauge], and at 5 inches of water differential pressure. This test console provides the pressure control and leakage measurement system required.

During astronaut transfer from the suit dressing room to the launching pad, a lightweight, hand-carried, portable ventilator provided suit cooling. Communications are maintained with the astronaut during this transfer by utilizing portable communication headsets carried by the astronaut insertion team.

In the MA-6 flight, the pressure suit served more as a flight suit since the cabin pressure was maintained. Astronaut comments indicated that the pressure suit was satisfactory throughout the flight.

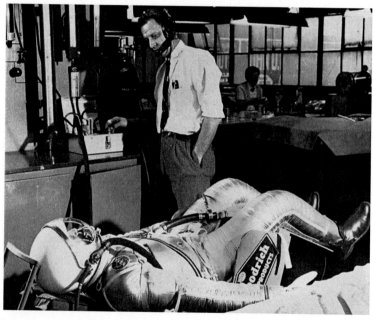

The pressure suit is inflated to check for any leaks. (Photo: Goodrich Corporation)

Appendix 6

Around the World with *Friendship 7*

A massive surge of interest in the space flight of John Glenn aboard Friendship 7 *convinced NASA to send the historic spacecraft on yet another flight around the world – this time at a far more sedate pace – allowing literally millions of people to view the vehicle that carried the first American astronaut to orbit the Earth. This was unofficially dubbed the spacecraft's "fourth orbit."*

The following contemporary reports following the global tour of Friendship 7 *were published in NASA's* Roundup Space News, *published bi-weekly by the Manned Spacecraft Center (now the Johnson Space Center) in Houston, Texas.*

NASA *SPACE NEWS ROUNDUP*, ISSUE MAY 2, 1962:

"*Friendship 7* Spacecraft Starts on World Tour; Schedule is Listed."
Friendship 7, the Mercury spacecraft in which Astronaut John Glenn Jr. orbited the Earth three times, is circling the Earth again. But this time it is making more than 20 stops along the way.

The National Aeronautics and Space Administration has loaned the spacecraft to the U.S. Information Agency which is displaying it on all continents. *Friendship 7* will return to the United States in mid-August for the Century 21 exposition at Seattle, Wash., before being presented to the Smithsonian Institute in Washington, D.C., for permanent exhibit.

The *Friendship 7* tour will include stops in Australia, Bermuda, Spain, Nigeria, Mexico, Great Britain, and Zanzibar – nations cooperating with the United States in the Project Mercury tracking program. The itinerary also includes two to four-day exhibitions in France, Japan, India, and Brazil.

The first stop on the trip was at Hamilton, Bermuda, April 20. Glenn was unable to accompany the spacecraft on the tour because he is assisting in preparations for forthcoming Mercury flights.

Cities in which the spacecraft will be seen, in addition to Hamilton are Accra, Africa; Ankara, Turkey; Bangkok, Thailand; Belgrade, Yugoslavia; Bogota, Colombia; Buenos Aires, Argentina; Cairo, Egypt; Djakarta, Indonesia; Karachi, Pakistan; Lagos, Nigeria; London,

Friendship 7 being unloaded at Manila Airport, Philippines, 20 July 1962. (Photo: NASA)

England; Madrid, Spain; Manila, Philippines; Mexico City, Mexico; New Delhi, India; Paris, France; Rio de Janeiro, Brazil; Santiago, Chile; Sydney, Australia; Tokyo, Japan; and Zanzibar.

NASA *SPACE NEWS ROUNDUP*, ISSUE JUNE 27, 1962:

"Glenn Spacecraft 'Orbits' World Taking More Time This Trip."

Friendship 7, the spacecraft in which Astronaut John Glenn made three rapid orbits of the Earth Feb. 20, is making quite a hit on its fourth world tour these days, although this one is being conducted at a much slower rate.

Beginning April 19 at Hamilton, Bermuda, the spacecraft has so far been in eighteen countries and today will make it twenty, when it is transferred from Bombay, W. India to Colombo, Ceylon.

Seven more countries are on the itinerary before *Friendship 7* returns to Seattle, Wash. and makes its final appearance.

NASA-MSC and U.S. Information Agency officials have been accompanying the spacecraft in relays to give lectures and answer questions at public appearances. With it in India and Ceylon at the moment is G. Merritt Preston, chief of Preflight Operations Division, and two USIA personnel.

From Ceylon, the spacecraft goes to Rangoon, Burma; Bangkok, Thailand; Djakarta, Indonesia and Sydney, Australia, accompanied by John J. Williams, of Preflight Operations,

Wherever *Friendship 7* traveled, it attracted massive crowds. (Photos: NASA)

then to Manila, Philippines; Tokyo, Japan; Seoul, Korea; and back to Seattle with Kenneth S. Kleinknecht, manager of Project Mercury.

Since mid-April in Bermuda, the spacecraft toured South America (Bogota, Colombia; Santiago, Chile; Buenos Aires, Argentina; Rio de Janeiro, Brazil; Mexico City) in the company of Arnold D. Aldrich, Flight Operations Division; Donald Gregory, technical assistant to the director; Chloe Wood, of the Office of Programs, NASA Headquarters and Richard S. Johnston, Assistant Chief of Life Systems Division.

Marshal Tito (closest to front), President of Yugoslavia (now Serbia) was given a personal inspection of *Friendship 7*. (Photo: Associated Press)

Friendship 7 arriving in England at RAF Bovingdon, Hertfordshire on 14 May 1962. (Photo courtesy of Mike Humphrey)

Of its two-day Mexico City stand, Gregory commented, "We turned them away in crowds at closing time. They stood in line for hours. Their only complaint was that we didn't give them long enough to look."

After a brief return to McDonnell Aircraft in St. Louis, Mo., for repairs and a half-day appearance at Dover, Del., the spacecraft went to Europe – to London, Paris, Belgrade, Yugoslavia, Madrid – then to Africa.

The line to view *Friendship 7* at London's Science Museum in South Kensington stretched all the way down Cromwell Road. (Photo: Science Museum, London)

Visitors in the Science Museum could climb a platform to view the spacecraft. (Photo: Science Museum, London)

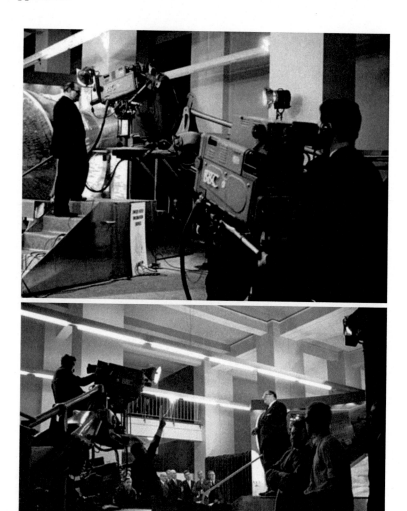

On the first evening of the *Friendship 7* display (14 May 1962) the BBC's *Panorama* presented a live broadcast from the Science Museum by host commentator Richard Dimbleby. (Photos: Science Museum, London)

An estimated 40,000 persons inspected the spacecraft during its three-day stay in London, 14–17 May, and according to Johnston the measure of interest shown was excellent. In addition to the general questions which were repeated in almost every city, Englishmen wanted to know why the capsule was black, what we learned from Glenn's flight, what problems Glenn had, and in addition asked detailed questions such as "What was the speed of the capsule when the drogue chute opened?"

Schoolboys, students, workmen in overalls with lunch boxes under their arms, businessmen carrying briefcases, youths in leather jackets and jeans, and elderly ladies with the morning's shopping in string bags thronged to the Science Museum for the display.

Snags of a strictly terrestrial nature sometimes upset the progress of the two-ton machine. Each of the capsule's orbits took about 88 minutes, but its progress from the RAF station at Bovingdon, Hertfordshire, where it was flown in by a U.S. Air Force Globemaster from the U.S., to the Science Museum in South Kensington where it was on display, took three hours. Part of the trouble was a loose wheel on the capsule trailer.

Said Johnston, "General reaction was excellent. The display is impressing a lot of people – doing us a lot of good." In France the language barrier furnished some difficulty not encountered in London, but write-ups in the French press indicated satisfaction.

In Madrid, Spain, 40,000 Spanish citizens viewed the display, forming a line nearly a mile long on May 27. The crowds were so heavy that the authorities had to call out mounted police to handle them, although there were no incidents.

Leaving Spain, the display toured through Accra, British W. Africa; Lagos and Kano, Nigeria; Cairo, Egypt; Istanbul, Turkey, and Karachi, Pakistan, before reaching Bombay last weekend. As you read this, it is probably in route to Colombo, Ceylon and the final month of overseas appearances before transportation back to the U.S. and Seattle.

NASA *SPACE NEWS ROUNDUP*, ISSUE JULY 25, 1962:

"*Friendship 7* Gets Rousing Reception."
G. Merritt Preston, Manager of Cape Operations and Chief of the Preflight Operations Division for NASA's Manned Spacecraft Center, returned recently from a tour of duty as NASA representative with the *Friendship 7* spacecraft.

People of many diverse cultures viewed the historic spacecraft up close. (Photo: NASA)

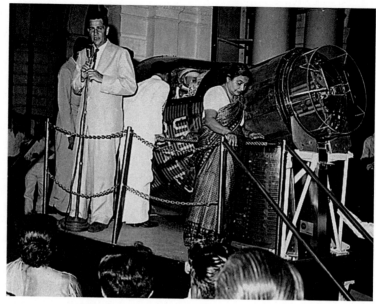

NASA's G. Merritt Preston was on hand to tell visiting dignitaries about the spacecraft and the MA-6 mission. (Photo: NASA)

The spacecraft which carried Astronaut John H. Glenn, Jr. on his history making flight is currently on the final legs of its world-wide tour.

Preston, who accompanied the spacecraft to Karachi, Pakistan; Bombay, India; Colombo, Ceylon; and Rangoon, Burma, was amazed at the reception given the event, the interest exhibited by the people and their apparent familiarity with the space flight.

He emphasized that much was accomplished by the tour in addition to showing the *Friendship 7* to the peoples of other countries. During his stay in the aforementioned countries, Preston was required to spend many working days which extended to 18 hours and longer – days which were made shorter by the realization of the good being done.

The highlight of the trip was the reception accorded the arrival of the exhibit at Bombay. At that city, more than one million people lined both sides of the street for 25 miles.

In Bombay, while the exhibit was open, the line to see the spacecraft was eight city blocks long and it took those waiting approximately five hours to get a six-second glance at the craft.

Preston admitted freely that there were frequent lighter moments along with the hard work on the strenuous trip. For instance, he found that in Karachi, the now-famous camel driver for Vice President Johnson holds a daily conference in front of the U.S. Embassy, at which time he grants interviews and signs autographs.

He also cited an incident in Rangoon which he classified as one of comic-terror. The iron gates surrounding the exhibit area were only opened about a foot in an endeavor to accommodate the crowd only in single file. The anxious crowd foiled this attempt in a mob scene as they crowded around the opening with such vigor that it was impossible for those in front to get through. Preston said, "they wound up to being three deep at the gate, some on their hands and knees, others standing on them, some holding babes in arms. Police inside the gate tried to assist by helping to pull those closest inside with a resultant loss of clothing in many instances. Another group of about 3,000 circumvented the problem by scaling the fence to gain admittance as they followed a small group of persons in distinctive garb 'over the top.' Once inside they straightened their clothing, smiled and were very polite."

Preston added that in Rangoon the exhibit was set up in a manner that two lines of persons could view the *Friendship 7* at the same time, one which would allow them to look into the capsule from close up, another which would permit a view from a raised platform and might really permit a better overall view of the exhibit. Long lines formed in both cases but when the people arrived at the foot of the stairs for the raised view they dropped out and went to the end of the other line in order that they might work their way forward and touch the spacecraft as they passed.

Among the interesting persons Preston met on the trip was Arthur Clark [*sic*], a successful British space science fiction writer. Clark had just returned to Ceylon from an American Rocket Society meeting in New York and had served as a panel moderator.

Since there were many demands upon Preston's time, a system was worked out to use local students in answering the many routine questions by those viewing the exhibit. At each stop a group of these students were briefed thoroughly and accepted the assignment with much enthusiasm. This permitted Preston to make appearances at universities and before important local groups to explain in greater detail the missions and objectives of America's space program.

In each case, Preston wrote to the students concerned and thanked them for their help during the exhibitions. These letters have had a profound impact and Preston is already receiving fan mail as a result. For example, Manzoor Ahmad of Karachi wrote, 'I am in receipt of your favor of June 22, 1962 appreciating my humble work in connection with the National Aeronautical Space Exhibition in Karachi. While I thank you for your kindness, I cannot help saying that whatever I did I did as a part of my duty to promote the cause of cooperation between the two countries and never deserve this high appreciation of yours. However, as I am a student of Hamia Talim-e-Milli College, Malir City, I value this letter and consider it a boon from heaven as I hope that this will improve my future career in life.'"

Preston re-emphasized that the value of the tour of the spacecraft and the accompanying good will which has resulted is inconceivable. He pointed out that among those observing the spacecraft during his stay with it were the local Communist leaders; leaders who looked at the *Friendship 7* and offered no comments.

#

In her September 2014 dissertation entitled *Project Apollo, Cold War Diplomacy and the American Framing of Global Interdependence*, researcher Teasel Muir-Harmony noted that the United States and the Soviet Union displayed their space accomplishments abroad in many different ways.[1]

"While the USIA designed exhibits and events to attract the largest crowds possible, and appeal to the general public around the world, the Soviet Union sent cosmonauts to foreign cities to meet with officials and heads of state.[2]

"Another major difference between American and Soviet space public diplomacy in the early 1960s had to do with the content of their space diplomacy events. While the United States sent space capsules around the world and broadcast launches and flights in real time, the Soviet Union organized cosmonaut tours and kept their space capsules and launches out of public view. Yuri Gagarin, the first human space traveler, toured a number of countries but his spacecraft was not put on public display; before 1965 only photographs of the Vostok veiled underneath a cover were shown to the world. And, historian Cathleen Lewis suggests that the lack of engineering information on the model of the first Vostok to be put on public display, 'Represented a deliberate effort to conceal the actual details of the human space-flight program in the Soviet Union by carefully camouflaging details about the design legacy of Vostok and its technical properties.'[3] When Glenn's capsule was put on display during its 'fourth orbit,' however, the USIA included engineering diagrams of its interior workings along with other exhibit components. This exhibit, as well as most American space exhibits in this period, highlighted scientific and technological advancements, as a demonstration of openness and a particular image of progress."[4]

In assessing the success of the spacecraft's global tour, John Glenn wrote to McGeorge Bundy, President Kennedy's National Security Advisor, explaining that the 'fourth orbit' of *Friendship 7* "stressed the fact that [the American space program] was not just a propaganda effort before the world, but a well-thought-out scientific program that could eventually benefit all peoples of the world as the scientific exploration it is." He went on to note that Russian exhibits highlighted personal appearances of cosmonauts while the United States emphasized scientific information via the capsule's display. According to Glenn, America's greatest advantage over the Soviet Union's space program was "the almost complete freedom to share experiences and new information."[5] He suggested that the openness of the American program, as represented by the display of the *Friendship 7* spacecraft, stood in for the nation and its political ideology – when the *Friendship 7* capsule was laid bare before the eyes of people from around the world it gave the impression that the United States' space program was real, benign, apolitical, and designed for the collective benefit of all mankind.[6]

At the end of the 24-nation global tour, *Friendship 7* had been seen by more than 4 million people, while another 20 million had watched television programs about the historic vehicle, broadcast from the different exhibition sites. The spacecraft was then placed on temporary display at the Century 21 exposition at Seattle, Washington, on 6 August 1962. The NASA exhibit was the only place that the historic space capsule would be displayed

In a ceremony held at the National Air Museum on 20 February 1963, on the anniversary of the flight of *Friendship 7*, NASA presented the spacecraft, John Glenn's spacesuit and other artifacts to the Smithsonian Institution. John Glenn is photographed speaking at the hand-over ceremony. (Photo: NASA)

in the United States before it was permanently moved to the Smithsonian Institution in Washington, D.C.

From 1963 to 1975 it was on display in the Smithsonian's Arts and Industries building in what was known until 1966 as the National Air Museum, and then the National Air and Space Museum. There was an interlude of some months when the Milestones of Flight building was under construction. This was opened on 1 July 1976, and *Friendship 7* took its rightful place next to the Wright brothers' original plane and Charles Lindbergh's *Spirit of St. Louis*. The historic spacecraft has occupied the same place in the central gallery of the museum ever since.

REFERENCES

1. Teasel Muir-Harmony, "Project Apollo, Cold War Diplomacy and the American Framing of Global Interdependence," submitted to the Program in Science, Technology, and Society in Partial Fulfilment of the Requirements for the Degree of Doctor of Philosophy in History, Anthropology, and Science, Technology and Society at the Massachusetts Institute of Technology, September 2014. Quoted with permission
2. Edward Murrow to Madrid USITO, 17 May 1962, Box 257, Folder "Outer Space, 14.B.5, Outer Space Exhibits, Jan-May, 1962, Part 2 of 2," Entry A1 3008-A, RG 59, NARA
3. Cathleen Lewis, "The birth of the Soviet space museums: creating the earthbound experience of space flight during the golden years of the Soviet space programme, 1957–68," in *Showcasing Space*, ed. Martin Collins (London: Science Museum, 2005) 142–158
4. Loyd Swenson Jr., James Grimwood, and Charles Alexander, *This New Ocean: A History of Project Mercury* (Washington, DC: National Aeronautics and Space Administration, 1966), 436; Donald Wilson to James Webb, 9 October 1962, Box 37, Entry A1 1039, RG 306, NARA
5. John Glenn, Jr. to McGeorge Bundy, 4 November 1963, Box 308, Folder "Space Activities, General 10/63-11/63," Papers of President Kennedy, National Security Files, Subject Files, John F. Kennedy Library
6. Teasel Muir-Harmony, "Project Apollo, Cold War Diplomacy and the American Framing of Global Interdependence," submitted to the Program in Science, Technology, and Society in Partial Fulfilment of the Requirements for the Degree of Doctor of Philosophy in History, Anthropology, and Science, Technology and Society at the Massachusetts Institute of Technology, September 2014. Quoted with permission.

About the author

Australian author Colin Burgess grew up in Sydney's southern suburbs. Initially working in the wages department of a major Sydney afternoon newspaper (where he first picked up his writing bug) and as a sales representative for a precious metals company, he subsequently joined Qantas Airways as a passenger handling agent in 1970 and two years later transferred to the airline's cabin crew. He would retire from Qantas as an onboard Customer Service Manager in 2002, after 32 years' service. During those flying years several of his books on the Australian prisoner-of-war experience and the first of his biographical books on space explorers such as Australian payload specialist Dr. Paul Scully-Power and Space Shuttle *Challenger* teacher Christa McAuliffe were published. He has also written extensively on spaceflight subjects for astronomy and space-related magazines in Australia, the United Kingdom, and the United States.

In 2003 the University of Nebraska Press appointed him series editor for the ongoing *Outward Odyssey* series of 12 books detailing the entire social history of space exploration, and he was involved in co-writing three of these volumes. His first Springer-Praxis book, *NASA's Scientist-Astronauts*, co-authored with British space historian David J. Shayler, was released in 2007. *Friendship 7* will be his eighth title with Springer-Praxis, for whom he is currently researching further books. He regularly attends astronaut functions in the United States and is well known to many of the pioneering space explorers, allowing him to conduct personal interviews for these books.

Colin and his wife Patricia still live just south of Sydney. They have two grown sons, two grandsons and a granddaughter.

© Springer International Publishing Switzerland 2015
C. Burgess, *Friendship 7*, Springer Praxis Books, DOI 10.1007/978-3-319-15654-5

Index

C. Burgess, *Friendship 7*, Springer Praxis Books, DOI 10.1007/978-3-319-15654-5

Printed in the United States
By Bookmasters